Henry Eeles Dresser

A Monograph of the Meropidæ or Family of the Bee-Eaters

Henry Eeles Dresser

A Monograph of the Meropidæ or Family of the Bee-Eaters

ISBN/EAN: 9783337201852

Printed in Europe, USA, Canada, Australia, Japan

Cover: Foto ©berggeist007 / pixelio.de

More available books at **www.hansebooks.com**

A MONOGRAPH

OF THE

MEROPIDÆ,

OR

FAMILY OF THE BEE-EATERS.

BY

H. E. DRESSER, F.L.S., F.Z.S., &c.,

MEMBER OF THE BRITISH ORNITHOLOGISTS' UNION, OF THE IMPERIAL SOCIETY OF NATURALISTS OF MOSCOW,
OF THE GERMAN ORNITHOLOGICAL SOCIETY, FOR. MEMB. OF THE AMERICAN ORNITHOLOGISTS'
UNION, CORR. MEMB. OF THE BOSTON SOCIETY OF NATURAL HISTORY, &c. &c.;
AUTHOR OF 'THE BIRDS OF EUROPE.'

Δεῖ σέ γ' ἄρ' ἀνθρώπων μερόπων γλυκερώτερον ᾔδειν,
ᾧ τε πέλει γλυκερὸν βρῶμα μέλισσα, μέροψ.

LONDON:
PUBLISHED BY THE AUTHOR AT 6 TENTERDEN STREET, W.
1884–1886.

ALERE FLAMMAM.

PRINTED BY TAYLOR AND FRANCIS,
RED LION COURT, FLEET STREET.

To the Memory

OF

ARTHUR HAY,

NINTH MARQUIS OF TWEEDDALE,

VISCOUNT WALDEN,

F.R.S., PRES. ZOOL. SOC., &c., &c.,

AS A SMALL TOKEN OF ESTEEM,

THIS WORK IS DEDICATED

BY

HIS SINCERE ADMIRER,

THE AUTHOR

CONTENTS.

LIST OF PLATES.

PREFACE.

It is now some years since I began to take a special interest in the Bee-eaters and Rollers (indeed long before I had completed my work on the 'Birds of Europe'), and I then commenced to make a collection of the birds belonging to these families, though not with the object of writing a Monograph of them, as I understood that my friend Mr. D. G. Elliot purposed doing so. However, as Mr. Elliot has not carried out what I understood to be his intention, and as I found that it would take years before I could gather sufficient material to commence a work on the birds of the Eastern Palæarctic Region, as a sequel to my 'Birds of Europe,' I resolved to undertake the present Monograph so as to employ my spare time in the meanwhile; and I trust that the result of my labours may be such as to merit the approbation of my fellow-ornithologists, and that the present work may prove the means of obtaining more information respecting several species of Bee-eaters about which, as will be seen by the following pages, but very little is known. I have to acknowledge with deep appreciation the willing assistance rendered to me by many fellow-naturalists during the progress of the present work; and I am especially indebted to Captain G. E. Shelley for the loan of his entire collection of *Meropidæ*, as also to Captain R. G. Wardlaw Ramsay for the loan of the Bee-eaters belonging to the Tweeddale collection. When I first determined to write this Monograph the late Mr. W. A. Forbes undertook to furnish the notes on the anatomy and osteology of the Bee-eaters and Rollers; but owing to his premature death, before he had done more than just commence the task, I was deprived of the valuable assistance he would thus have given me. However, his successor in the post of Prosector to the Zoological Society, Mr. Frank E. Beddard, has most kindly come forward and volunteered to undertake this portion of the work, and to him I am indebted for the valuable notes on this subject that are embodied in the Introduction. I may add that I have now in preparation a Monograph of the Rollers, which I trust ere long to complete (and, indeed, all the Plates have already been drawn on stone, printed off, and coloured), which will form a companion volume to the present Monograph.

H. E. DRESSER.

Topclyffe Grange, Farnborough, Kent.
28 December, 1885.

INTRODUCTION.

GENERAL REMARKS.

THE Bee-eaters must certainly be ranged next to the Rollers, to which they are very closely allied, and they are also nearly allied to the Jacamars, as also, but in a less degree, to the Kingfishers, Motmots, Hoopers, and Hornbills. On the whole, I should be inclined to range the Meropidæ between the Coraciidæ and the Galbulidæ; and this view coincides with those expressed by Mr. Frank E. Beddard, to whom I am indebted for the following notes :—

"In endeavouring to determine the systematic position of the Meropidæ it is only necessary to take into consideration the Coccygomorphæ, Celeomorphæ, and Coracomorphæ of Prof. Huxley, which together nearly correspond to the Anomalogonatæ of Mr. Garrod. No author, so far as I am aware, has indicated any special affinities of the Meropidæ with families outside of this group of birds.

"Accordingly, in the following notes upon the anatomy of the Meropidæ, the only comparisons made are with the other genera of the Anomalogonatæ.

"EXTERNAL CHARACTERS AND PTERYLOSIS.

"The oil-gland of *Merops* is nude, and in this it resembles the Passeres, as well as the Coraciidæ, Galbulidæ, and Trogonidæ. The rectrices are, as Nitzsch states, 12 in number. An aftershaft is present, as in *Coracias* and several allied genera.

"The pterylosis of *Merops* has been described by Nitzsch; the following is a more detailed account :—

"The spinal tract is wide and closely feathered; at first it is perfectly continuous round the neck with the ventral tract, but about halfway down the neck becomes separate. It is narrow and closely feathered on the neck, and terminates in a truncated or sometimes bifurcate extremity between the scapulæ, about halfway between the two extremities of the bone; at this point there is a distinct break, and the rest of the spinal tract is double, composed of two slender bands enclosing a widish space over the spine. The form of each of the two posterior halves of the spinal tract is conical, the base of the cone being directed forwards; posteriorly each tract narrows to the width of two feathers, and the two tracts fuse a short way in front of the oil-gland. Anteriorly the two divisions of the spinal tract are connected by scattered feathering with the humeral tract. The ventral tract is double from close to its point of origin at the mandibular symphysis ; the two tracts are at first very narrow, and separated from each other and from the sides of the head by spaces without any feathering : on the pectoral region each ventral tract is of uniform width, four or five feathers wide ; below the axilla each tract gives off a closely feathered lateral branch to the hypopteron. The pectoral tract of each side dwindles to a single row of feathers before ending just in front of the cloaca.

"The pterylosis of *Merops* does not differ in any important respects from that of *Coracias*, which it appears most closely to resemble ; it differs from *Momotus* and *Todus* in having a double spinal tract, and from *Galbula* in having no inner branch to the pectoral tract running along the clavicles.

"OSTEOLOGY.

"The characters of the skull of *Merops* have been dwelt upon by Prof. Huxley, in his well-known paper on the Classification of Birds. The skull is desmognathous, and the family is placed in the group Coccygomorphæ, in close proximity to the Alcedinidæ, Buccrotidæ, Upupidæ, Momotidæ, and Coraciidæ. In the same paper the author refers to the shape of the sternum in the Coccygomorphæ, which is stated to present usually two notches on each side, and to be devoid of a bifurcated manubrial process, the one exception to the last statement being *Merops*. I find, however, that *Trogon* (*T. reinhardti*) has a distinctly bifurcate manubrial process, though not perhaps so well marked as in *Merops*. The characters of the skull of the former genus, which has been shown by Mr. Forbes to be schizognathous and not desmognathous, would appear to necessitate its removal altogether from Huxley's group Coccygomorphæ, though it undoubtedly agrees in its visceral and muscular anatomy, as well as in external characters, with the other families of this order, and presents affinities to *Merops*, as well as to the Celeomorphæ and Coracomorphæ, by the possession of a forked manubrium sterni.

"The sternum of *Merops* is, as stated by Prof. Huxley, provided with two lateral notches, of which the outer one on each side is the deepest ; in *Coracias*, *Trogon*, *Pteroglossus*, *Bucco*, *Dacelo*, and *Yunx* the sternum is closely similar ; in *Momotus* the notches are converted by continuous growth into foramina ; in *Upupa* and *Buceros* there is only a single pair of notches.

"According to Prof. Huxley the clavicles of the Coccygomorphæ are without any process developed backwards from the summit of their 'symphysis.' This statement is true, in so far that there is no junction between such a process and the sternum as exists in many Passeres, not to mention other orders of birds ; but the process itself exists, in a rudimentary form, in many of the Coccygomorphæ. In *Coracias* and *Trogon* there is such a process developed just at the junction of the two clavicles, and there is the faintest trace of a similar structure in *Merops*. On the other hand, *Upupa*, *Yunx*, *Bucco*, *Momotus*, &c. have no median process of the furcula.

"MYOLOGY.

"The Meropidæ agree with all the other families of the order Anomalogonatæ of Prof. Garrod in the absence of the *ambiens*, and in the presence of the *femoro-caudal* and the *semitendinosus* and its accessory; the formula, therefore, on Prof. Garrod's system, is—A X Y. With regard to other muscles which vary in different groups of birds, the Meropidæ possess the so-called *expansor secundariorum*, which has the characteristic disposition termed 'Ciconiine' by Prof. Garrod. The presence of this muscle is not usual among Anomalogonatous birds, but it exists among the Momotidæ, Alcedinidæ, in *Steatornis*, the Coraciidæ, Leptosomidæ, and Galbulidæ; in the last three families the *expansor secundariorum* is Ciconiine, as in *Merops*.

"The *tensor patagii longus* arises from the clavicle by two distinct heads, one in common with the single head of the *tensor patagii brevis*; the tendon of the latter muscle bifurcates just before its insertion on to the *extensor metacarpi*. The termination of the *tensor patagii* tendon is, in fact, very closely similar to that of the Galbulidæ, especially *Urogalba paradisea*; in *Coracias* the tendons of this muscle are a little more complicated.

"The *deltoid* extends a long way down the humerus; it receives a fibrous cord from the scapula, which passes under the *latissimus dorsi* and over the *anconeus longus*. A similar accessory head to the deltoid occurs in many birds—for instance, among those which have evident relations to the Bee-eaters, in *Momotus*, *Hylomanes*, *Galbula albirostris*, and *Urogalba paradisea*; in *Coracias*, as in the Todies, this tendinous slip is absent. The *anconeus longus* is attached to the humerus about one third of the way down by a tendinous slip, which is inserted in common with the posterior *latissimus dorsi*; a similar disposition is met with in *Coracias*, *Urogalba*, and *Galbula*; this tendinous accessory head of the *anconeus* is absent, however, in the Todies, and in *Momotus*, *Hylomanes*, *Dacelo*, &c. The deep flexor tendons have been described by Garrod as more particularly resembling those of *Momotus* and *Dacelo*; they differ slightly from *Coracias*, and still more from the Galbulidæ.

"On the whole the myology of the Bee-eaters seems to indicate a specially close alliance with the Coraciidæ and Galbulidæ, though there are no wide differences from other families of the Anomalogonatæ. The presence of cæca in the intestine, and the absence of a tuft on the oil-gland, led Prof. Garrod to associate the Meropidæ with the above-named families, as well as with the Trogonidæ and Passeres. The absence of cæca and the presence of a *tufted* oil-gland distinguish the Hornbills, Colies, Kingfishers, Woodpeckers, and Motmots. The myology of these different families does not perhaps afford any very strong support to Prof. Garrod's division, but it is at any rate in no way opposed to it: it will be noticed from the facts recorded above that the *expansor secundariorum*, if present, has a different disposition in the Aves Passeriformes such as *Merops*; the accessory head of the *anconeus* is apparently absent in the Trogonidæ, and often, if not always, present in the Passeres. The myology as well as other details of the anatomy of the Caprimulgidæ are not perhaps in accord with such a division; but it appears to me that there are grounds for removing this group altogether from the Piciformes or Passeriformes."

HISTORY OF THE BEE-EATERS.

Owing to its conspicuous coloration and to its being a common bird in Southern Europe, the Bee-eater was well known to the early writers on natural history; but it is scarcely necessary to trace its history beyond Brisson, who, in 1760, treated of all the then known species under the one generic title *Apiaster*, which, however, according to our present rules of nomenclature, has to give place to *Merops* of Linnæus. Brisson placed his genus *Apiaster* near the Kingfishers, but separated it from them by the Todies, and ranged it between *Todus* and *Buceros*, which latter he terms *Hydrocorax*. Thirteen species are enumerated by Brisson under his genus *Apiaster*, of which, however, only seven can be included as true Bee-eaters, viz.:—Nos. 1 and 2, *Merops apiaster*; No. 6, *Merops bicolor*; No. 7, *Merops superciliosus*; Nos. 8 and 9, *Merops viridis*; No. 11, *Melittophagus pusillus*; No. 12, *Merops philippinus*; and No. 13, *Melittophagus quinticolor*. Nos. 3 and 10 are doubtful, and Nos. 4 and 5 certainly not Bee-eaters.

Linnæus, in the 12th edition of his Syst. Nat. (1766), places the genus *Merops* between *Alcedo* (in which he included the Jacamars) and *Upupa* (in which *Promerops* is included). He only includes seven species; but these must be reduced again to three, viz. *Merops apiaster* (Nos. 1 and 3), *Merops viridis* (No. 2), and *Merops superciliosus* (No. 4). Some years later D'Aubenton, in the 'Planches Enluminées,' figured two species of Bee-eaters not previously known, viz. Pl. Enl. 252 (*Merops bicolor*) and Pl. Enl. 649 (*Merops nubicus*), but neither of these received scientific names till some years later.

In 1773 Pallas, in the Supplement to his 'Reise im russischen Reichs,' first described and named *Merops persicus*; and in 1776 P. L. S. Müller, in the Supplement to his 'Natursystem,' discriminated *Melittophagus pusillus*, and at the same time also gave the specific name of *americanus* to the species figured by D'Aubenton in the 'Planches Enluminées' (no. 252), which name, however, being inapplicable, has to be rejected in favour of *bicolor* of Boddaert, given in 1783 (Table des Planches Enluminées d'Histoire Naturelle de M. D'Aubenton).

J. G. Gmelin, in 1788, in the 13th ed. of Linnæus's Syst. Nat., enumerates the species previously described, together with several which are not referable to the present family, and describes *Merops nubicus*, which species is founded on D'Aubenton's plate No. 649 in the 'Planches Enluminées.' He also describes (p. 463), under the name of *Merops erythrocephalus*, a Bee-eater which may probably be *Melittophagus quinticolor*; but the description is not sufficiently clear to enable his name to be used.

Latham (Ind. Orn.) and Bonnaterre (Tabl. Encycl. et Méth.) both gave a *résumé*, in 1790, of the species previously described; but neither described any new species or added anything worthy of note to the literature of this family. In 1793, however, Lichtenstein, in a pamphlet, now of great rarity (Cat. rer. nat. rariss.), first described the Swallow-tailed Bee-eater, *Dicrocercus hirundineus*; and though his description is very meagre, yet it is quite clear enough to show that it is referable to this species. In Shaw and Nodder's 'Naturalist's Miscellany' (1790–1813) Shaw describes *Melittophagus gularis* and *Merops malimbicus*, and the following species are figured, viz. *Merops nubicus* (pls. 78, 613), *Merops apiaster* (pl. 102), *Melittophagus gularis* (pl. 337), and *Merops malimbicus* (pl. 701). On pl. 357 an illustration of a Bee-eater is given which is not recognizable, and to which he gives the name of *Merops erythrocephalus*; it bears some resemblance to *Melittophagus quinticolor*, but cannot be recognized as that species.

Latham, in the Supplement to his Ind. Orn., described *Merops ornatus* in 1801; and from then we pass over a period of five years, without finding anything worthy of note, to 1807, when Levaillant published his beautiful work ' Histoire Naturelle des Promérops et des Guêpiers,' in which twenty-one folio illustrations of the Bee-eaters were issued with accompanying letter-press, but no scientific names were given. All these plates are readily recognizable, except in the case of Levaillant's *Guêpier adansou* (pl. 13), which is depicted as somewhat resembling *Merops bicolor*, but having a red tail; and, so far as I can judge, the plate must have been drawn from a made-up skin, as no such bird has ever been found by subsequent ornithologists. In 1817, Vieillot (Nouv. Dict. d'Hist. Nat. vol. xiv.) gave the specific names of *albicollis* to Levaillant's *Guêpier à gorge blanche ou le Guêpier Cuvier* (pl. 9), of *quinticolor* (*Melittophagus quinticolor*) to his *Guêpier quinticolor* (pl. 15), of *leschenaulti* (*Melittophagus leschenaulti*) to his *Guêpier laichenot* (pl. 18), and of *bulocki* (*Melittophagus bullocki*) to his *Guêpier à gorge rouge ou le Guêpier bulock* (pl. 20).

Passing over a space of four years, we find *Merops sumatranus* described by Raffles in 1821 (Trans. Linn. Soc. xiii. p. 294), and in 1824 Temminck (Pl. Col. no. 310) figured and described *Nyctiornis amictus*.

Until 1828 the Bee-eaters were all included in one and the same genus, that of *Merops*, but in that year Boie (Isis, 1828, p. 316) proposed the generic title of *Melittophagus* for the small Bee-eaters which have the tail even, lacking the elongation on the central rectrices ; and this genus is one that I consider should stand; and in the same article Boie also gave the specific name of *sonnini* to *Melittophagus sonninii*. In 1829 (Ill. Orn. ii. pl. 58) Jardine and Selby figured and described *Nyctiornis athertoni* ; and in 1831–32, Swainson (Zool. Ill. 2nd ser. vol. ii.) pro-posed the generic title of *Nyctiornis* for the two large square-tailed Bee-eaters having the elongated pectoral plumes (*N. amictus* and *N. atherloni*), and in the same volume he also gave a second generic title, *Nyctinomus*, for the same species, Isidore Geoffroy St.-Hilaire, about the same time, in 1832 (Nouv. Ann. du Mus. d'Hist. Nat. i.), proposing the title of *Alcemerops* for this small group ; whereas in 1836 (J. As. Soc. Beng. v. p. 360) Hodgson proposed the generic name of *Bucia*, and again, in 1841 (J. As. Soc. Beng. x. p. 29), that of *Napophila*, for the same group. In 1834, Smith (S. Afr. Quart. Journ. 2nd ser. part ii. p. 320) described *Melittophagus bullockoides*, which species was also figured by him in the Ill. Zool. S. Afr., Aves, pl. ix.

Guérin in 1843 (Rev. Zool. 1843, p. 322) first gave the generic title of *lafresnayei* to *Melitto-phagus lafresnayei*, which species was figured in 1847 in the Atlas to Ferret and Galinier's 'Voyage en Abyssinie,' pl. 15.

In 1846 Des Murs (Rev. Zool. 1846, p. 243) separated the Carmine-throated Bee-eater from *Merops nubicus*, with which it had up to then been united, giving the specific name of *nubicoides*, which it still retains, to the former. I may here point out a mistake in my synonymy of this species, for I have inadvertently placed amongst the list of titles that of *Merops superbus*, Vieillot (Nouv. Dict. xiv. p. 23, 1817), which is a synonym of *Merops nubicus* and not of *M. nubicoides*.

In 1849 Sundevall (Œfv. K. Vet.-Ak. Förh. 1849, p. 162) proposed to separate *Melittophagus gularis* from its allies, and to give it the generic title of *Meropiscus* ; but I cannot agree with him in so doing, and consider that it ought to be retained amongst those species grouped together under the genus *Melittophagus*. Bonaparte, in his 'Conspectus generum Avium,' i. p. 164, published in 1850 a description of Forsten's Bee-eater, which until then bore only the MS. name, given to it by Temminck, of *forsteni*, which name Bonaparte also retained, and at the same time very rightly separated it generically from its allies, bestowing on it the title of *Meropogon*.

In 1852 Reichenbach ('Handbuch der speciellen Ornithologie,' Meropinæ, pp. 61–83) gave a monographic sketch of the family, with descriptions of all the then known species of Bee-eaters, accompanied by coloured illustrations; but the illustrations are poor, and the letterpress is merely a compilation from the writings of previous authors and of but little scientific value. In the earlier portion of the article he subdivides the Bee-eaters into three genera—*Merops*, *Meropogon*, and *Nyctiornis*; but at p. 82 he gives a review of the Bee-eaters, in which he places them in four sections, viz. :—i. *Melittotherinæ*, containing three genera, *Melittotheres* (*Merops nubicus* and *M. nubicoides*), *Tephraërops* (*Merops malimbicus*), and *Melittophas* (*Merops bicolor*); ii. *Apiastrinæ*, containing four genera, *Aërops* (*Merops albicollis*), *Merops* (*Merops apiaster* and *Melittophagus leschenaulti*), *Sphecophobus* (*Melittophagus pusillus*, *M. sonninii*, and *M. lafresnayei*), and *Melittophagus* (*Dicrocercus hirundineus* and *Merops ornatus*); iii. *Phlothrinæ*, containing two genera, *Phlothrus* (*Merops viridis*) and *Blepharomerops* (*Merops persicus* and *M. philippinus*); and iv. *Nyctiorninæ*, containing five genera, *Meropiscus* (*Melittophagus gularis*), *Coccolarynx* (*Melittophagus bullocki* and *M. bullockoides*), *Meropogon* (*M. forsteni*), *Nyctiornis* (*N. amictus*), and *Bucia* (*N. athertoni*). From the above it will be seen that the generic divisions are somewhat erratic and peculiar. At p. 75 he attempts to resuscitate, under the name of *Merops adansoni*, Levaillant's *Guêpier adanson*, and figures it pl. cccexlviii. 3243, copying Levaillant's plate. This supposed species is, I may here mention, the *Merops senegalensis* of Shaw (Gen. Zool. viii. pt. i. p. 163, 1812) and the *M. longicauda* of Vieillot (Nouv. Dict. d'Hist. Nat. xiv. p. 15, 1817). In 1854 Bonaparte (Consp. Volucr. Anisod. p. 8) proposed the generic title of *Urica*, his type being *Melittophagus quinticolor*; but this lapses into a synonym of *Melittophagus*. Cassin, in 1857 (Journ. Ac. Sc. Phil. 1857, p. 37), described and figured the still so little known *Melittophagus muelleri*, under the name of *Meropiscus muelleri*; and in 1858 Von Pelzeln (Sitzungsb. k. Ak. Wiss. Wien, xxxi. p. 320) described *Melittophagus boleslavskii*. In 1859 Cabanis (Mus. Hein. ii. pp. 133–138) differentiated the following five genera, viz. *Sphecouax* (type *Melittophagus bullockoides*), *Melittias* (type *Melittophagus leschenaulti*), *Dicrocercus* (type *D. hirundineus*), *Cosmaërops* (type *Merops ornatus*), and *Pogonomerops* (*Meropogon forsteni*), of which only one, *Dicrocercus*, will stand. At the same time he described, under the name of *Phlothrus cyanophrys*, the Blue-throated Green Bee-eater (*Merops cyanophrys*), which proves to be an excellent species.

The same year (1859), Cassin (Proc. Acad. Nat. Sci. Phil. 1859, p. 14) described the very distinct *Merops breweri*, first calling it by that name, but later (op. cit. p. 34) referring to it the generic title of *Meropogon*; and the same year Heine (J. f. O. 1859, p. 434) proposed for it a new generic title, viz. that of *Bombylonax*, whereas one year later Hartlaub (in Wiegm. Archiv, xxvi. p. 90) proposed another generic name, viz. that of *Archimerops*, for the same species. This is the last generic title that I find on record for any of the Bee-eaters. As will be seen from the above notes, twenty-four generic names have been proposed for the Bee-eaters, of which I can only adopt five, viz. *Nyctiornis*, *Meropogon*, *Merops*, *Dicrocercus*, and *Melittophagus*. Cabanis in 1869 (Von der Decken's Reisen in Ost-Afr. iii. p. 34) described as new a species under the name of *Merops cyanostictus*; but, so far as I can judge, it is a very slightly marked form of *Melittophagus pusillus*, and not worthy of specific rank.

From this date to 1882 I find no new genera proposed and no new species described; but in that year a very distinct new species, *Merops boehmi*, was described by Dr. Reichenow (Orn. Centralblatt, April 1882), and in the following year (Revoil's Faune et Flore Çomalis, Ois. p. 5, 1883) M. Oustalet described and figured *Melittophagus revoili*, a very excellent and distinct species, which will be found figured in the present work. Since then one more species of Bee-eater (*Merops muscatensis*) has been differentiated by my friend and late colleague Mr. R. Bowdler

Sharpe, which, though closely allied to *Merops cyanophrys*, and to extreme forms of *Merops viridis*, and forming, as it were, a link between the two species, is a fairly recognizable and separable species. During the progress of the present work several articles on the different species belonging to the present family have been issued, amongst which I may cite that from the pen of Dr. Reichenow (J. f. O. 1885, p. 222), who, in pointing out slight differences remarked by him between examples of *Melittophagus gularis* from northern and southern localities, says :—" In the typical form from the Gold Coast, as also in specimens from Liberia, the forehead and a broad superciliary stripe are pale cobalt like the rump. Individuals from Angola and the Congo, however, have the frontal line blue-green and the superciliary stripe olive-green and but indistinctly defined. Individuals from the Gaboon and Cameroons agree with the Angolan examples, except that the superciliary stripe is more distinct and blue-green." This southern race he proposes to call *Melittophagus gularis australis*, and adds that he defers the question as to whether there is a third race of *Melittophagus gularis*, and for the present only proposes to separate the northern and southern races, the northern limit of the range of the southern race being, he thinks, the Cameroons. In the Niger district he believes that the typical race alone occurs.

I have not had an opportunity of examining specimens of this Bee-eater from Angola ; but I have now before me two examples from Gaboon which differ in no respect from some of the specimens I have in my collection from Fantee, and I have also other specimens from the Gold Coast which have the frontal stripe very narrow and tinged with green, and the superciliary line but very indistinctly defined and dull in colour, and it therefore appears to me that in all probability the Angolan specimens described by Dr. Reichenow will prove to be but younger examples of true *Melittophagus gularis*.

Still more recently, indeed during the time I have been engaged on the final part of the present work, my friend Prof. Wilhelm Blasius, in a most painstaking article on the avifauna of Celebes (Madarász, Zeitschr. für gesammte Ornithol. iii. p. 239, 1885), proposes to give subspecific rank to a form of *Merops philippinus* inhabiting Celebes, stating his reasons *in extenso*, and he proposes to call it *Merops philippinus*, var. *celebensis*. This form he considers distinct on account of its having the back and head darker and more olivaceous than in typical *M. philippinus*, the rufous colour on the throat not being sharply separated from the olivaceous brown of the breast, but gradually merging into it, and in having the blue coloration on the abdomen less clearly defined. It appears, however, that both this form and typical *M. philippinus* are found together in India, Ceylon, British Burmah, &c., and that the two forms run into each other so much that I cannot see any just reason to separate them even subspecifically. Further than this, I have before me a specimen from Celebes which does not at all agree with Prof. Blasius's description, and is undistinguishable from typical *Merops philippinus* ; and the specimen I have figured, which was obtained in Ceylon by Mr. Holdsworth, and received by me from him direct (hence there can be no doubt as to the precise locality where it was obtained), agrees most closely with Prof. Blasius's description of his var. *celebensis*. For these reasons I do not think that his bird is deserving even of subspecific rank.

CLASSIFICATION.

The Bee-eaters differ but little *inter se* in internal structure and in pterylosis, except that, as stated by Mr. W. A. Forbes (Monogr. of Jacamars, p. xi, footnote), most of the Meropidæ have only the left carotid, whereas *Nyctiornis* has two, and it is therefore necessary to define the genera by external characters. This family has been greatly subdivided by various authors, and has been split up into as many as twenty-four genera; but of these it appears to me advisable to recognize only five, viz.:—

1. *Nyctiornis*, which has the tail square and the pectoral plumes much elongated.
2. *Meropogon*, which has the pectoral plumes as in *Nyctiornis*, but has the middle rectrices elongated as in *Merops*.
3. *Merops*, which has the pectoral plumes not elongated, but has the two middle rectrices much elongated.
4. *Dicrocercus*, which has the pectoral plumes not elongated, but has the tail deeply forked.
5. *Melittophagus*, which also has the pectoral plumes not elongated, but has the tail nearly or quite even.

Full particulars of the characters of each genus are given in the body of the present work, so that I need not recapitulate them here; and as the osteology has already been amply illustrated, and as there are no special peculiarities to figure in the soft parts, I have not deemed it necessary to issue any plate to illustrate the osteology or the generic characters.

HABITS AND DISTRIBUTION.

All the particulars I have been able to glean respecting the habits of the birds belonging to the present family will be found in the body of the present work. They are inhabitants of the forests and plains, as a rule affecting localities close to rivers, and are arboreal in their habits. Most of the species are gregarious even during the breeding-season and nest in colonies; but the two species belonging to the genus *Nyctiornis* are said to be much less gregarious in their habits than their allies, and are usually seen singly or in pairs. The note of all the Bee-eaters is said to be harsh and unmelodious, and, as a rule, they are silent birds.

All the Bee-eaters the nesting-habits of which are known make their nest-holes, which they themselves excavate in the ground, usually in banks, and most frequently in those which skirt or are near rivers or streams. When I wrote the articles on *Nyctiornis amictus* and *N. athertoni* in the present work, as there stated, nothing certain was known respecting the nidification of these two birds, and it was uncertain whether they nested in hollow trees or in holes in banks. Since then, however, the question has been satisfactorily solved; for in a letter lately received from Mr. W. Davison he says:—"It will doubtless interest you to know that Morgan took two nests of *Nyctiornis athertoni* last year, in October and November. The bird breeds, like other Bee-eaters, in holes in banks. The holes are made by the birds themselves and extend six or seven feet into the bank. In one case the bird had nested in the bank of a road, in the other in an old elephant-pit."

All the Bee-eaters without exception lay pure white, glossy, roundish eggs.

The range of this Family is confined to the Old World, none of its members occurring in

America, where it may perhaps be considered to be replaced by the Galbulidæ. It is represented in the Palæarctic, Ethiopian, Indian, and Australian Regions, greatly predominating, however, in the second. The Palæarctic Region is really inhabited by only *four* species, though a single example of a fifth is recorded as having occurred (but so far from its usual range as to justify its being regarded as only an accidental and even a doubtful straggler). The Ethiopian Region, including therein Southern Arabia, is inhabited by no fewer than *twenty-one* out of the *thirty-one* species recognized in this volume; and of these *eighteen*, or considerably more than half of the whole number, are peculiar to that Region. In the Indian Region *eleven* species are met with, of which *four* are peculiar to its continental portion, and those are common to the Palæarctic and Ethiopian Regions, while *two* are peculiar to its islands. The Australian Region includes only *three* species—*one* widely distributed, the other *two* restricted to the remarkable island of Celebes, one of them forming the type of a distinct genus, *Meropogon*. Of the other genera, *Nyctiornis* is confined to the Indian Region, *Dicrocercus* to the Ethiopian, and *Melittophagus* inhabits both in common; while *Merops* has more or fewer representatives in all four Regions.

The precise distribution of the Bee-eaters may be more plainly exhibited by the subjoined Table (p. xx).

TABLE OF THE GEOGRAPHICAL DISTRIBUTION OF THE MEROPIDÆ.

	Palæarctic Region.									Ethiopian Region.					Indian Region.							Australian Region.				
	Northern Europe.	Central Europe.	Southern Europe.	North Egypt and Barbary.	Asia Minor.	Palestine.	Northern Arabia.	Persia.	Central Asia.	North-east African and Southern Arabia.	East Africa.	West Africa.	South Africa.	Madagascar.	India and Burmah.	China and Cochin China.	Malay Peninsula.	Andaman.	Java.	Borneo.	Philippines.	Celebes.	Moluccan Group.	Papuan Group.	Timor Group.	Australia.
Nyctiornis amictus															•	•	•		•	•		•				
athertoni															•	•										
Meropogon forsteni																						•				
Merops breweri											•	•														
swainsonii												•														
bicolor															•	•	•	•	•	•	•					
viridis					•	•	•	•	•	•					•	•	•	•	•	•	•					
muscatensis								•		•																
cyanophrys										•																
boehmi											•															
albicollis										•	•	•														
ornatus																						•	•	•	•	•
philippinus †	•	•	•	•	•	•	•	•	•	•	•				•	•	•	•	•	•	•		•			
persicus			•	•	•	•	•	•	•	•	•	•			•	•										
superciliosus				•						•	•	•														
apiaster		•	•	•	•	•	•	•	•	•	•	•	•		•											
malimbicus											•	•														
nubicus				•		•	•			•	•	•	•													
nubicoides											•		•													
Dicrocercus hirundineus											•	•	•													
Melittophagus lafresnayei											•	•														
sommii												•														
pusillus										•	•	•	•													
quinticolor															•		•		•	•						
leschenaulti															•	•	•		•	•						
gularis												•														
muelleri												•														
bullockoides											•	•	•													
bullocki											•	•														
boldcalavskii												•														
revoili													•													

† Has been once recorded as having straggled to the British Isles.

Family MEROPIDÆ.

Genus NYCTIORNIS.

Merops, apud Jardine & Selby, Ill. Orn. pl. 58 (1829), nec Linn.
Nyctiornis, Swainson, Zool. Ill. 2nd ser. vol. ii. pl. 56 (1831–32). Type *Merops amictus*, Temm. Pl. Col. 310.
Nyctinomus, Swainson, ut suprà (1831–32). Type *M. amictus*, ut suprà.
Alcemerops, Isid. Geoffr. St.-Hil. Nouv. Ann. du Mus. d'Hist. Nat. i. p. 395 (1832). Type *M. amictus*, ut suprà.
Bucia, Hodgson, J. As. Soc. Beng. v. p. 360 (1836). Type *Merops athertoni*, J. & S. Ill. Orn. ii. pl. 58.
Napophila, Hodgson, op. cit. x. p. 29 (1841). Type *M. athertoni*, ut suprà.

HAB. India; Burmah and the Malay peninsula, down to Sumatra and Borneo.

Alis brevibus, rotundatis; remige primâ brevissimâ, quartâ omnium longissimâ, tertiâ vix breviore: caudâ vix emarginatâ: rostro longo, curvato, robusto: juguli plumis elongatis, attenuatis: pedibus brevibus, robustis.

Bill long, curved, pointed; culmen curved, flattened and grooved laterally from the base for some distance, compressed towards the tip, which is acute. Nostrils basal, lateral, rounded, covered with short stiff feathers. Tarsi short, stout, anteriorly scutellate. Toes long, the lateral ones slightly unequal; hind toe long, stout, padded beneath. Wings short, broad, but reaching beyond the base of the tail; first quill short, somewhat shorter than the secondaries; the third and fourth longest, the latter slightly longer than the third. Tail long, broad, somewhat emarginate. Feathers on the crown slightly elongated, those on the throat much more so, somewhat attenuated towards the tip.—Type *Nyctiornis amictus*.

THIS genus contains only two species, *Nyctiornis amictus* and *Nyctiornis athertoni*, which in general habits do not differ much from the species included in the genus *Merops*. They frequent the vicinity of woods, feed on insects, which they capture chiefly on the wing, and, like their allies, are frequently to be seen seated on a dead bough or in some open place, from whence they sally forth to capture their prey. Nothing certain is known respecting their nidification; but they are supposed to nest in holes in trees or banks and to deposit pure white glossy eggs.

PINK-CROWNED BEE EATER.

RYNCHORNIS AMICTUS.

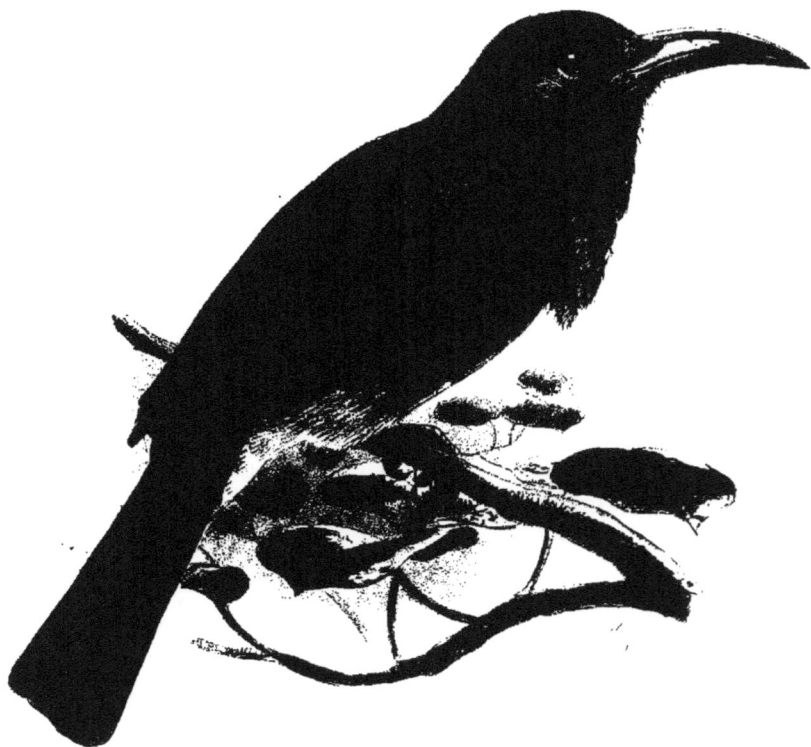

FINE CROWNED BEE-EATER

NYCTIORNIS AMICTUS.

PINK-CROWNED BEE-EATER.

Merops amictus, Temminck, Pl. Col. livr. 52, no. 310 (1824) ; Lesson, Traité d'Orn. p. 237 (1831) ; Schlegel, Mus. Pays-Bas, *Merops*, p. 13 (1863).

Nyctiornis amictus (Temm.), Swains. Zool. Ill. 2nd ser. ii. pl. 56 (1831) ; id. Classif. of B. ii. p. 333 (1837) ; Eyton, P. Z. S. 1839, p. 101 ; Gray, Gen. of B. i. p. 87 (1846) ; Bp. Consp. Gen. Av. i. p. 164 (1850) ; Gould, Birds of Asia, vol. iii. pl. (1850) ; Licht. Nomencl. Av. p. 66 (1854) ; Cab. & Heine, Mus. Hein. ii. p. 133 (1859) ; Gray, Hand-l. of B. i. p. 98, no. 1194 (1869) ; Hume, Str. Feath. viii. pp. 48, 85 (1879), Bingham, Str. Feath. viii. p. 193 (1879) ; Kelham, Ibis, 1881, p. 378.

Alcemerops amicta (Temm.), Blyth, Cat. B. Mus. As. Soc. p. 51 (1849).

Alcemerops amicta (Temm.), Hartl. Verz. Mus. Brem. p. 15.

Nyctiornis amicta (Temm.), Moore, P. Z. S. 1854, p. 264 ; Sclater, P. Z. S. 1863, p. 214 ; Beavan, Ibis, 1869, p. 408 ; Salvadori, Ucc. Born. p. 91 (1874) ; Tweeddale, Ibis, 1877, p. 298 ; Sharpe, Ibis, 1877, p. 6 ; Hume & Davison, Str. Feath. vi. p. 498 (1878) ; Sharpe, Ibis, 1879, p. 248 ; Bingham, Str. Feath. ix. p. 153 (1880).

Nyctiornis malaccensis, Cab. & Heine, Mus. Hein. ii. p. 133 (1859–60) ; Gray, Hand-l. of B. i. p. 98. no. 1195 (1869) ; Hume, Str. Feath. ii. p. 469 (1874).

Figuræ notabiles.

Temminck, Pl. Col. 310 ; Swainson, Zool. Ill. 2nd ser. ii. pl. 56 ; Gould, B. of Asia, iii. pl.

Hab. Burmah ; Malay peninsula ; Sumatra, Borneo.

Ad. corpore suprà, alis et caudâ psittacino-viridibus : pileo antico rosaceo-lilacino, plumis nasalibus, genis anticis mentoque summo cyaneo-viridibus : genis reliquis miniatis, posticè lilacinis : gutture toto et juguli plumis miniatis elongatis, his medialiter nigricanti-viridibus : gulæ jugulique plumis elongatis, rubris : alis subtùs ochraceo-cervinis : caudâ subtùs flavâ nigro terminatâ, rectrice extimâ externè nigrâ : corpore subtùs reliquo viridi, plumis medialiter pallidè viridibus : rostro nigro : pedibus plumbeis : iride rubrâ.

Juv. adulto similis, sed gulâ, fronte et jugulo viridibus.

Adult male (Malacca).—Upper parts generally rich deep parrot-green ; fore part of the crown to about the centre of the head of a beautiful lilac-pink colour ; wings and tail rather darker than the back ; quills with the inner margin and tip blackish, the shafts black ; feathers covering the nostrils and a small spot on the chin and base of cheeks pale greenish blue ; remainder of chin and the elongated feathers on the throat deep vermilion-red, the latter with dark green centres ; ear-coverts and sides of neck of the same colour as the back ; rest of underparts pale green, with greenish-white centres to the feathers ; under surface of wings at the base pale ochreous buff ; under surface of tail yellow, broadly terminated with black, and the outer feather on each side externally black ; bill black, greenish at base of lower mandible ; legs bluish lead-grey ; iris red. Total length about 10 inches, culmen 2·2, wing 5·0, tail 4·75, tarsus 0·6.

Adult female (Malacca).—Resembles the male, but is smaller, and differs in having the forehead and lores of the same colour as the throat, the mauve-pink on the forehead being restricted to a somewhat narrow band; the red on the breast is also narrower.

Young (Lampong, Sumatra).—Differs from the adult in having the crown and throat green instead of red, and the iris greyish brown.

According to Davison, the adult bird has the legs and feet pale green, often dingy, sometimes bluish; bill black, whity brown at base from nostril to gape of upper mandible, and lower mandible from about angle of genys to base; irides bright yellow to orange-yellow; eyelids dark plumbeous green.

Obs. In very old birds the outer webs of most of the quills are edged with bluish green, and in one specimen from Malacca this is the case to a very perceptible degree; but in other examples from Malacca and Sumatra, where it occurs, it is developed only to a slight extent. The sexes do not differ in coloration, except as above stated, but the female is somewhat smaller than the male.

THE range of the present species, one of the largest as well as one of the most richly coloured of the Bee-eaters, extends from British Burmah down the Malay peninsula to Sumatra and Borneo; but it does not occur in India, being there replaced by its congener *Nyctiornis athertoni.* Capt. Bingham met with it near Moulmein, and remarks that whereas he elsewhere only met with it in pairs, he there observed a small flock. In Tenasserim, he informs me, it is a rare bird in such of the forests as he visited. He first procured it at Kya-en on the Zammee river, about 16° N. lat. He also ('Stray Feathers,' ix. p. 153) "shot a male on the 19th November, 1879, in the Thoungyeen valley, near the banks of the Thingaugyeonoun choung, a feeder of the Meplay."

On the Malay peninsula it appears to be generally distributed, and collections from Malacca usually contain examples of this beautiful bird. Lieut. Kelham obtained two alive, at Kwala Kangsar, Perak, on the 28th of February; and he adds that he does not believe it to be a common bird. I have examined many examples from Sumatra, where, apparently, it is by no means rare; and in Borneo it is said to be tolerably common in some localities. Mr. Treacher obtained it on the Lawas river; and Mr. Mottley (P. Z. S. 1863, p. 214) says that it is "rather common at Gunong Tabok, on the Riam Kanan river, but I suppose rare elsewhere. My hunter says that it is not known far in the interior." The immature bird, lacking the pink frontal patch and the red throat, has been described from Malacca as a distinct species under the name of *Nyctiornis malaccensis*; but a specimen in this plain green dress has a single pink feather on the forehead, showing that it is a young bird commencing to assume the adult plumage; and I can fully indorse the opinion expressed by Count Salvadori and the late Lord Tweeddale, that the Malaccan bird does not in any respect differ from the species found in Tenasserim, Burmah, and Borneo. In measurement examples from Malacca vary as follows—wing 4·7 to 5·15, tail 4·5 to 5·0; from Sumatra—wing 4·85 to 5·15, tail 4·7 to 5·10; and a female from Sarawak measures—wing 4·7, tail 4·7.

This richly coloured bird is said to inhabit the forests, but to avoid the denser portions, frequenting those parts where the large trees are somewhat scattered and where the sunlight can penetrate, and its favourite haunts are the banks of large streams and borders of swamps surrounded by forest. It is, unlike most of the Bee-eaters, said to be solitary in its habits, being seldom seen in flocks, and is not shy, but stupid, and confiding rather than timorous. Capt. Bingham informs me that he met with it in Tenasserim, where it affected the denser forests, and was solitary in its habits, being usually seen in pairs. It is, he adds, "a very silent bird, and once only have I heard its note, which is a deep croak ending in a guttural *k-r-r-r*. This was on the Kaukarit and Meeawuddy road, when a pair came and sat on a branch of a tree not ten feet above my head. One of them, the male as it proved to be, when I had shot it, several times uttered its croak, bobbing its head at each utterance and swelling out its pectoral plumes. It is a stupid bird, not easily frightened, as I found in the above case when I shot the male first with one barrel, and the female afterwards with the second, though they were seated scarcely three feet apart." Mr. Mottley says (*l. c.*) that "its note is something between the croak of a frog and the 'churr' of a Fern-Owl, often repeated and sustained perhaps half a minute."

Lieut. Kelham obtained two alive and put them into his aviary, where at first they did well, feeding on plantains and hopping about most cheerfully, every now and again flirting up their long tails after the manner of *Copsychus musicus*; but after a few days they sickened and died. These birds have, he adds, a peculiar and most aromatic smell about them.

Mr. W. Davison states ('Stray Feathers,' vi. p. 69) that the most northern point in Tenasserim where he saw and obtained this bird in the plains was at a village four days' march south of Yea, about 14° 30′ N. lat., and adds that from this point it gets less uncommon as one goes south. In the hills, however, it extends further north, and on the slopes of Mooleyit he got it in nearly 17° N. lat. "This species," he writes, "less often occurs away from the forest than *Nyctiornis athertoni*; but although keeping, as a rule, to the woods, it avoids the denser portions, frequenting those parts where the larger trees are somewhat scattered, and where plenty of sunlight penetrates; favourite places are the banks of large streams and the borders of swamps and shallow lagoons surrounded by forest.

"The note of this bird is somewhat similar to that of *Nyctiornis athertoni*, and is a hoarse *quo-quă-quă-quă*, uttered at irregular intervals. When one calls it is usually answered by its mate, the birds being generally found in pairs, seldom singly, and never, that I know of, in parties. When uttering its note the bird leans forward, stretches out its neck, and puffs out the feathers of its throat, and at each syllable of its note bobs its head up and down.

"It breeds, I should say, about March and April, as on the 20th of March I shot a female, out of which I took an egg that was fully formed, but still quite soft; but I was unable to find the nest.

"I have not noticed that either this bird or *Nyctiornis athertoni* were crepuscular. Occasionally on a clear moonlight evening, about seven or eight o'clock, I have heard their note; but there are numbers of birds that of a bright evening, or if they have been in any way disturbed, will call. Like the true Bee-eaters it lives entirely on insects, which it takes on the wing."

So far as I can ascertain, nothing definite is known respecting the breeding-habits of this bird. Capt. Bingham believes that it breeds in holes in the ground, like the other Bee-eaters, and about the end of April or the beginning of May; for the pair he shot on the 28th of April showed on dissection that they were breeding. I think, however, that they will be found to breed in hollow trees, and, in all probability, the eggs are pure white. Mr. Everett writes ('Ibis,' 1877, p. 6) that "a nest containing two eggs was brought me at Belidah in January. The eggs were rather small in comparison with the size of the bird, nearly equal at both ends, and spotted with faint red in a ring round the larger end, the ground being white. The nest was neatly lined with dry grass inside, and exteriorly was roughly put together with bamboo-leaves and rush." I give these details as given by Mr. Everett, but at the same time must say that I feel convinced that he was deceived, and that the present species does not lay spotted eggs, nor build a nest as described by him.

In the preparation of the above article I have examined the following specimens :—

E Mus. H. E. Dresser.

a, ♂. Malacca, 1877 (*H. Kelham*). b, ♀; c, ad.; d, juv. Malacca.

E Mus. Brit.

a. Tenasserim (*Packman*). b, ♂. Tenasserim, 8th December, 1870 (*E. W. Oates*). c. Singapore. d. Malacca (*Harvey*). e. Malacca, April 1854 (*Cantor*). f, ♂. Malacca, 1854 (*A. R. Wallace*). g. Penang (*Gould coll.*). h. Sumatra. i, ♂. Sumatra, 1861 (*A. R. Wallace*). k. Borneo. l. Busan, Borneo (*A. Everett*). m. Borneo (*Sir J. Brooke*). n. Labuan (*H. Low*). o, juv. Tagorn, Sarawak, May 1875 (*Everett*).

E Mus. Tweeddale.

a, b, c, d, e, f, ad.; g, juv. Malacca, 1873 and 1874 (*Wardlaw Ramsay*). h, ♀. Mount Ophir, Malacca, 11th August 1873 (*W. Ramsay*). i, k. Sumatra (*Bock*). l, ♂; m, ♂; n, ♂; o, ♀. Sumatra, September 1878 (*Bock*). p, q, ♂. Sumatra, October 1878 (*Bock*). r, juv. Lampong, Sumatra (*Bock*). s, ♀. Sarawak, Borneo, 10th January, 1870.

E Mus. Paris.

a, b, ad.; c, juv. Sumatra (*Duvaucel*). d. Borneo (*Toussaint*).

BLUE BACKED BEE EATER
NYCTIORNIS ATHERTONI

NYCTIORNIS ATHERTONI.

BLUE-NECKED BEE-EATER.

Merops athertoni, Jardine & Selby, Ill. Orn. ii. pl. 58 (1820) ; Schlegel, Mus. Pays-Bas, *Merops*, p. 12 (1863).
Bucia nipalensis, Hodgs. Journ. As. Soc. Beng. v. p. 361 (1830) ; Blyth, J. As. Soc. Beng. x. p. 922 (1842).
Nyctiornis cæruleus, Swains. Classif. of B. ii. p. 333 (1837).
Nyctiornis amherstiana, Royle, Ill. Himal. Bot. i. p. lxxvii (1839).
Merops cyanogularis, Jerdon, Madr. Journ. xi. p. 229 (1840).
Alcemerops paleazureus, Less. Rev. Zool. 1840, p. 262.
Bucia athertoni, Blyth, J. As. Soc. Beng. x. p. 922 (1841).
Nyctiornis athertoni (Jard. & Selby), M'Clelland, P. Z. S. 1839, p. 155; Gray, Gen. of B. i. p. 87 (1849) ; Bp.
    Consp. Gen. Av. i. p. 164 (1850) ; Gould, B. of Asia, vol. iii. pl. (1850) ; Licht. Nomencl. Av. p. 66 (1854) ;
    Horsf. & Moore, Cat. B. E. Ind. Co. Mus. i. p. 89 (1854) ; Cab. & Heine, Mus. Hein. ii. p. 132 (1859) ;
    Jerdon, B. of India, i. p. 211 (1862) ; Gray, Cat. B. Nepal, p. 58 (1863) ; Blyth, Ibis, 1866, p. 345 ; Beavan,
    Ibis, 1869, p. 107 ; Blanf. Ibis, 1870, p. 465 ; Hume, Nests & Eggs of Ind. B. p. 108 (1873) ; Fairbank,
    Stray Feathers, v. p. 394 (1877) ; Hume & Davison, Str. Feath. vi. p. 68 (1878) ; Anderson, Zool. Exped.
    Yunnan, i. p. 583 (1878) ; Tiraut, Bull. C. A. Coch. Ch. scr. 3, vol. i. p. 98 (1879) ; Bingham, Str. Feath.
    ix. p. 153 (1880) ; Butler, Str. Feath. ix. p. 382 (1880) ; Oates, B. of Brit. Burm. ii. p. 63 (1883).
Napophila athertoni (Jard. & Selby), Blyth, Journ. As. Soc. Beng. xi. p. 104 (1842).
Napophila meropina, Hodgs. in Gray's Zool. Misc. p. 82 (1844).
Alcemerops athertoni (Jard. & Selby), Blyth, Cat. B. in Mus. As. Soc. p. 52 (1849).
Bucia athertoni (Jard. & Selby), Reichenb. Handb. Meropinæ, p. 83 (1852) ; Gray, Hand-l. of B. i. p. 98, no. 1196
    (1869).
Merops assamensis, M'Clell. fide Gray, Gen. of B. i. p. 87 (1849).

Figuræ notabiles.

Jardine & Selby, *l. c.*; Gould, B. of Asia, vol. iii. pl. ; Reichenbach, *l. c.*

HAB. India, ranging eastward into Burmah.

Ad. capite, collo et corpore suprà cum alis caudâque psittacino-viridibus : fronte pallidè turcino-cæruleâ et pileo
eodem colore lavato : gulæ et juguli plumis elongatis, turcino-cæruleis, imis saturatiùs cæruleis : pectore imo
et corpore subtùs reliquo ochraceo-cervinis, abdomine saturatè viridi striato : alis subtùs aurantiaco-cervinis :
rostro saturatè brunneo : mandibulâ ad basin pallidè schistaceâ : pedibus pallidè viridibus : iride aurantiaco-
rubrâ seu aurantiaco-fuscâ.

Juv. saturatior : capite, gulâ et jugulo viridibus nec cæruleis.

Adult male (Tenasserim, 10th June).—Head, neck, and upper parts generally rich parrot-
green ; forehead pale turquoise-blue, and the crown marked with the same colour ; a broad line

from the chin extending down the breast rich turquoise-blue, varying to deep ultramarine-blue; lower breast and underparts generally ochreous buff, the abdomen broadly striped with dark green; vent and under tail-coverts plain buff; under surface of the wings warm orange-buff, that of the tail yellowish buff, tipped with brown; bill dark horn, light at the base of the lower mandible; legs pale dingy green; claws horn-colour; iris orange-red or orange-brown. Total length about 13 inches, culmen 1·8, wing 5·6, tail 5·7, tarsus 0·75.

Adult female (Tenasserim, 20th Oct.).—Resembles the male, but is smaller, measuring—Total length 11·5 inches, culmen 1·3, wing 5·25, tail 5·3, tarsus 0·7.

Young.—Differs from the adult in having the forehead and throat dark green and not blue, and the general tinge of green in the plumage is darker than in the adult.

THIS, the largest of the Bee-eaters, inhabits the Indian region only, being found in India proper, whence it ranges eastward into Burmah and Cochin China. Dr. Jerdon (B. of India, i. p. 212) says that "it is found in the large and lofty forests of India. I have found it in Malabar in several localities; well up the sides of the Neilgherries at least to 3000 feet; in the Wynaad jungles; and, on one occasion, on the Naekenary Pass, leading from the Carnatic into Mysore, at about 1400 feet. It does not appear to occur in Central India, but is not rare all along the Himalayan range from the Deyra Doon to Assam, Arrakan, and Tenasserim. I got it at Darjeeling at about 4000 feet high." According to Capt. Butler (Str. Feath. ix. p. 382) it is "rare in the Deccan and Southern Mahratta country. It was obtained by Laird in the forest tract west of Belgaum, but he did not hear of any other instance of its occurrence within the region." Mr. Davison informs me that it is by no means a common species in Southern India, but has been obtained at Kulhutty in the Neilgherries, at Villayar in Coimbatore, and at Carcoor and Nelambore in Malabar. Dr. Fairbank, during a visit to the Palani hills in Southern India, obtained this bird, in December 1866, at the head of the Kamban valley, which skirts the Palanis along the south-eastern base, and he also observed a pair at Periûr on the lower Palanis in March. Hodgson procured it in Nepal and at Darjeeling, and there are examples in the British Museum from Travancore and Nynee-Tal. It is found in Burmah, Tenasserim, Siam, and Cochin China; and Mr. Oates writes (B. of Brit. Burmah, ii. p. 64) that "it occurs sparingly throughout British Burmah, frequenting forest country and being a constant resident. I procured a few specimens in the Arrakan hills near Nyoungyo, and I met with it near Pegu town once or twice. Mr. Blanford got it at Bassein, and Capt. Wardlaw Ramsay at Tonghoo and on the Karin hills." Mr. Hume records it from Pahpoor, Amherst, Karope, and Tavoy in Tenasserim; Blyth says that in the Southern Tenasserim provinces it occurs together with *Nyctiornis amictus.* Messrs. Hume and Davison write ('Stray Feathers,' vi. p. 68) that "this Bee-eater is found in Tenasserim at Pahpoor, Thatone, the Salween river, Thoungsha, the Gyne river, Kanee, Khyin, and Amherst, and is sparingly distributed throughout the better-wooded and less elevated parts of the northern and central parts of the province;" and Mr. Davison says that he "met with it only at Amherst and northwards of that place; it is nowhere very common and only occurred in the better wooded portions of the country. As a rule it prefers to keep to the forest, but it occasionally wanders into gardens, and at Amherst I shot two specimens off a large peepul-tree growing some considerable distance from any forest. I have not met with this species south of Amherst, though it possibly does occur somewhat further south." Capt. Bingham observed it near Kaukarit, and adds that it was far commoner along the well-wooded Dawna range and in the Thoungyeen valley. Dr. Anderson also obtained it near Bhamo.

In the British Museum there is a specimen collected by Mouhot in Cambodia; and Dr. Tiraut writes (l. c.) that it "inhabits the open portions of all the large forests of Cochin China : I have killed it at Dông-lâch, at Cái-Cung, Suôi-nuoc, and Srok-tranh. It carries itself like the other Bee-eaters, but lives in solitude. I kept one in confinement for more than three weeks. It had a broken wing and could not fly far. Usually it sat still and attentive on its perch, uttering a harsh cry like a Roller when offered a live insect."

In habits the present species does not appreciably differ from the members of the genus *Merops*. It frequents wooded localities, but is far from being a denizen of the dense forests like *Nyctiornis amictus*. Mr. Davison informs me that it is "a silent and rather shy bird, rarely uttering a cry except when alarmed. It frequents the borders of the dense forests of Southern India and may occasionally be seen perched on a bamboo overhanging some stream on the look out for insects. They are almost invariably found in pairs. The Irulars or Hillmen declare that they breed in holes in river-banks; but I have never succeeded in obtaining the eggs myself." Captain Beavan, who procured it at Moulmein, says that it was seated on a dead bough of a tree which overhung a tank, from which it sallied forth every few minutes like the ordinary Bee-eater and returned to its perch with an insect. It allowed him to approach within easy shot without seeming at all annoyed by his presence.

Hodgson, writing on the habits of this species, says (Journ. As. Soc. Beng. v. p. 361) that "they are of rare occurrence and are solitary woodlanders. They are found in the lower and central regions of Nepal, but seldom or never in the northern. Their food consists of bees and their congeners, but they likewise consume great quantities of scarabæi and their like; they seek the deep recesses of the forests, and there, tranquilly seated on a high tree, watch the casual advent of their prey, and having seized it, return directly to their station. They are of dull staid manners, and never quit the deepest recesses of the forest. In the rajah's shooting-excursions they are frequently taken alive by the clamorous multitude of sportsmen, some two or more of whom single out a bird, and presently make him captive, disconcerted as he is by the noise. The intestinal canal in this bird is usually about twelve inches long, with cæca of an inch or more in length placed near to the bottom of it. The stomach is muscular and of medial subequal thickness. Such, too, is the character of the stomach and intestines in *Merops*."

I am also indebted to Capt. Bingham for the following notes :—

"This species is fairly common in the Tenasserim forests, affecting the more open forests and gardens ; I have even procured it on some solitary trees in a paddy-field near Kaukarit on the Houndraw river. It has a hoarse chuckling croak, and if a pair happen to be seated near each other when one takes a flight the other croaks, bending its head down and puffing out its pectoral plumes, ending up with a long ' k-r-r-r.' Another note it has by which it has frequently betrayed its vicinity to me, which is not at all unlike the self-satisfied short ' cluk ' of *Rhopodytes tristis*. This bird is a veritable bee-eater; out of the many I have shot, I have rarely found one whose stomach was not crammed with bees.

" Lieut. Atherton, the discoverer of this species, informed Selby that it was ' very scarce and rare, inhabiting the thickest jungles in the interior of India, and feeding by night, at which time it was very noisy, repeating frequently the short cry of *curr, curr*.' Thus it will be seen that the information he obtained respecting the habits of this Bee-eater is considerably at variance with what I have gathered from later collectors.

"The following is a note I sent about its breeding to 'Stray Feathers':—

"'I cannot positively vouch for the four eggs said to belong to this species which I have procured. The case stands thus :—On the 23rd April a Karen named Myat-jo, in my employ, brought me four roundish, white, very glossy eggs, and the dead body of a bird of this species, which on dissection proved to be a female evidently breeding. His story was that he had watched the bird go into a hole in the sandy bank of the Meplay stream, and dug it out, catching it alive seated on the four eggs he had brought me. As the place was not more than a mile or so from the place where I had pitched my camp, I went off at once with him to inspect the spot. Examination of the ruined nest and further questioning of Myat-jo elicited the following :— A tunnel had been dug by the birds into the soft bank to a depth of seven or eight feet, ending in a rounded chamber. The eggs reposed on the bare ground, there being no attempt at a nest. The bird pecked vigorously at Myat-jo's hand, when from time to time he put it in to ascertain how much further he had to dig. The eggs were very hard-set, and I had much difficulty in cleaning them out; they measure—$1·13 \times 1·05$, $1·16 \times 1·02$, $1·12 \times 1·04$, and $1·17 \times 1·02$.

"'Myat-jo being an aboriginal Karen, and belonging to a village to which missionaries have not yet penetrated, I myself have little doubt that the eggs are authentic. I have, moreover, never yet found him trying to impose on me.'"

To this Mr. Hume appends the following remarks (Stray Feath. ix. p. 472) :—"On the whole I am inclined to accept the eggs. There is no doubt that they are undistinguishable from the eggs of *Halcyon smyrnensis*; but nevertheless there are several reasons for believing that they may belong to *N. athertoni*. In the first place, I have never known *Halcyon smyrnensis* bore anything like so deep a tunnel. In the second place, the female specimen of *N. athertoni* said to have been caught on the eggs proves to be a female that had been recently laying. It had been caught and not shot; and if he did not catch it in the hole, it is difficult to understand how the Karen could have got hold of it. In the third place, the eggs are precisely what the bird might have been expected to lay.

"At the same time it must be admitted that we have hitherto had reason to suppose that this bird bred in holes in trees, and Captain Bingham himself once shot a breeding bird issuing from such a hole, and very few species of birds lay both in holes of trees and in holes in sandy banks."

The specimen figured is in my own collection.

In the preparation of the above article I have examined the following specimens :—

E Mus. H. E. Dresser.

a, ♂ ad. Kaukarit, Tenasserim, 10th June, 1879 (*C. T. Bingham*). *b*, ♂. Thoungyeen valley, Tenasserim, 23rd August, 1879 (*C. T. Bingham*). *c*, ♀. Thoungyeen Forests, Tenasserim, 20th October, 1879 (*C. T. Bingham*). *d*. Himalayas (*Gerrard*).

E Mus. Tweeddale.

a. Deyra Doon. *b*. Nynee-Tal (*Pinwill*). *c*. Burmah. *d*. Tonghoo. *e*, ♂. Tonghoo, 3rd November, 1874 (*Wardlaw Ramsay*). *f*, ♂. Tonghoo, 10th November, 1875 (*W. R.*). *g*, ♀. Tonghoo, 22nd October, 1874 (*W. R.*). *h, i, j, k, l*. Karin hills, September and October, 1874 (*W. R.*). *m*, ♂. Moulmein, October 1865. *n*. Assam. *o*, ♂; *p*, ♀. Khasia hills, January 1876 (*Chennell*).

11

E Mus. Brit.

a, *b*, *c*, *d*. Nepal (*Hodgson*). *e*. Behar (*Hodgson*). *f*. Darjeeling (*Hodgson*). *g*. Nynec-Tal (*Pinwill*). *h*. Travancore. *i*. Burmah (*Gould coll.*). *k*. Bhamo, Burmah, February 1868 (*Dr. Anderson*). *l*, ♀ ad. Kaukarit, Tenasserim (*Bingham*). *m*, ♀. Arrakan hills, January 1872 (*E. W. Oates*). *n*. Siam, August 1868 (*E. Pierre*). *o*. Cambodia (*Mouhot*).

E Mus. Acad. Cantabr.

a. (*Coll. Selby ex Atherton*) Typus.

c 3

Genus MEROPOGON.

Meropogon, Bp. Consp. Gen. Av. i. p. 164 (1850).   Type *Merops forsteni*, Temm. in Mus. Lugd.

Pogonomerops, Cab. & Heine, Mus. Hein. ii. p. 132 (1859).   Type ut suprà.

Nyctiornis, Giebel, Thes. Orn. ii. p. 733 (1875), nec Swainson.

Hab. Celebes.

Alis et juguli plumis elongatis sicut in *Nyctiornithe*: caudâ cum rectricibus elongatis sicut in *Merope*: rostro elongato, curvato, robusto: pedibus brevibus, robustis.

The present genus contains but a single species, which is closely allied to both *Nyctiornis* and *Merops*, having the elongated pectoral plumes of *Nyctiornis* and the elongated central rectrices of *Merops*, and forms a true link between the two genera.   The late Marquis of Tweeddale published (Trans. Zool. Soc. viii. p. 111) some excellent notes respecting the affinities of *M. forsteni*, which I cannot do better than transcribe :—"This species has the first primary half the length of the second, which is a little shorter than the third.   The third and fourth are longest, and equal.   The fifth is somewhat shorter than the third and fourth, but longer than the second.   In the structure of the wing, therefore, it differs from both *Merops* and *Melittophagus*, but agrees with *Nyctiornis*.   The grooved culmen of *Nyctiornis* is not present; but a shallow channel extends from the base of the maxilla, on both sides of the culmen, for two thirds of its length.   This character is not possessed by either *Nyctiornis*, *Merops*, or *Melittophagus*.   The rectrices are truncated, as in *Nyctiornis*; but the middle pair are elongated, as in *Merops*, and closely resemble in form and proportion those of *M. philippensis*.   The feet are those of the family.   The elongated pectoral plumes resemble in character the same feathers in *Nyctiornis*.   Altogether *M. forsteni* may be regarded as a link uniting *Nyctiornis* to *Merops*, but most nearly allied to *Nyctiornis*."

The single species belonging to this genus is found only in Celebes, all particulars as yet known respecting it being given in the following article.

SUMPTUOUS BEE-EATER
MALE AND FEMALE

MEROPOGON FORSTENI.

FORSTEN'S BEE-EATER.

Merops forsteni, Temminck in Mus. Lugd.; Schlegel, Mus. Pays-Bas, *Merops*, p. 8 (1863); Meyer, J. f. Orn. 1871, p. 231.

Meropogon forsteni, Bp. Consp. Gen. Av. i. p. 164 (1850, ex Temm. MSS.); Gray, Hand-l. of B. i. p. 98, no. 1199 (1869); Meyer, J. f. O. 1872, p. 405; Tweeddale, Trans. Zool. Soc. viii. p. 42 (1872); Gould, B. of Asia, part xxv. (1873); Meyer, Ibis, 1879, p. 58.

Pogonomerops forsteni (Bp.), Cab. & Heine, Mus. Hein. ii. p. 132, footnote (1859-60).

Nyctiornis forsteni (Bp.), Giebel, Thes. Orn. ii. p. 733 (1875).

Figura unica.

Gould, B. of Asia, part xxv.

Hab. Celebes.

♂ *ad.* pileo, capitis lateribus et jugulo cum abdomine intensè azureis : corpore suprà, alis et rectricibus centralibus elongatis, saturatè psittacino-viridibus : remigibus primariis iu pogonio interno sordidè nigro marginatis et eodem colore apicatis : rectricibus (centralibus exceptis) castaneis, externis vix viridi marginatis : abdomine imo sordidè nigro : subcaudalibus sordidè castaneis : rostro et pedibus nigricantibus : iride coccineâ.

♀ similis, sed sordidior.

Adult (Rurukan).—Fore part of the crown, forehead, sides of the head, and entire throat, breast, and fore part of the abdomen rich deep glossy cobalt-blue ; hinder crown, nape, and neck blackish brown ; entire upper parts with the two central elongated rectrices deep glossy parrot-green ; wings deep green, the shafts of the feathers black, the primaries internally margined and tipped with dull black ; tail (except the two central rectrices) deep fox-red, the outer feather on each side with the outer web dark dull green ; under wing-coverts white, except the border of the wing, which is dull green ; lower abdomen dull blackish brown ; under tail-coverts dull coppery brown, marked with green ; bill black ; legs dull black ; iris crimson. Total length about 12 inches, culmen 2·1, wing 4·5, tail 7·1, tarsus 0·45, central rectrices extending 2·55 inches beyond the lateral ones.

Adult female.—Similar to the male, but duller in colour.

THE present species, one of the rarest of the Bee-eaters, is, so far as we know, confined solely to Celebes. For a long time the specimen at Leyden was the only one known to exist, and this remained unique until other examples were obtained by Dr. Meyer in the locality where Forsten first discovered it.

In a letter written from Menado, Celebes, in March 1871, Dr. A. B. Meyer gives (J. f. O. 1871, p. 231) the following facts respecting this bird :—" In 1840 Forsten discovered a bird which Schlegel subsequently described under the name of *Merops forsteni*. Only one specimen was sent to Leyden, and it has long been a desideratum with ornithologists, both on account of its rarity and more especially because of its resemblance to another, a West-African, species. In Leyden every exertion was made to obtain further specimens of *M. forsteni*, but without avail. Wallace did not succeed in finding it; and Rosenberg remained some time at the place where it was first obtained in order to procure it, but was unable to secure a single specimen. I have now succeeded in obtaining several examples, both males and females, at the very place near Rurukan where Wallace collected for some time. This lovely bird lives in the dense forests in places difficult of access, is found on the highest trees, and in habits resembles the other Meropidæ. It does not appear to be at all rare, but is difficult to find, as it retires to the dense forests. Thus the bird is unknown to the natives, and I only succeeded in procuring it after infinite trouble."

Beyond this all that we know respecting this species is found in Dr. A. B. Meyer's further notes published in 1879 (' Ibis,' 1879, pp. 58, 59), as follows :—" There existed before my journey to Celebes only one male specimen of this interesting species in the Leyden Museum, obtained by Forsten, in the year 1840, near Tondano, at an elevation of 2000 feet in the Minahassa. Professor Schlegel showed me the specimen before I went away in 1870, and urged me to rediscover it, as none of Forsten's successors (Wallace, Rosenberg, and others) had brought it home. Mr. Wallace, in his charming book, ' The Malay Archipelago ' (i. p. 429), says, in the chapter on the ' Natural History of Celebes,' ' In the next family, the Bee-eaters, is another equally isolated bird, *Meropogon forsteni*, which combines the characters of African and Indian Bee-eaters, and whose only near ally, *Meropogon breweri*, was discovered by M. Du Chaillu in West Africa !' African affinities being said to give a characteristic feature to the Celebean fauna, and, besides, *M. forsteni* being so rare that the Celebean origin of the bird was doubted, I resolved to do my best in searching after it. I therefore made about a hundred coloured sketches, and distributed them among the natives, to send away into the mountainous districts, and put a relatively high reward on a skin. I got the first specimen at the end of the month of May 1871 from a forest near Rurukan, not very far from the place where Forsten had procured his specimen some thirty years before; and afterwards, in June, I found the bird in the richest virgin forest which I have seen in these regions, on the way from Langowan (about 2000 feet) to Pangku, where it appeared to be not so rare. I suppose that *M. forsteni* only inhabits the mountainous districts, like *Euodes erythrophrys*, *Hemiphaga forsteni*, &c.; but, of course, I am not sure of this. I should not say these birds are rare, but only known to occur in restricted localities ; if only these localities are discovered, the bird proves then to be numerous. It is the same with certain butterflies which have been declared to be rare ones, such as *Papilio blumei*, *P. androcles*, &c. ; they also do not, or at least rarely, occur near Menado, where most travellers have collected, and therefore have the reputation of being rare ; but I found places in Celebes where any quantity of them can be procured. They are not collected in greater quantities because nearly every one who travels there does not remain a long time on those spots. It will be the same with other animals. Of course there are also animals which really are only represented by very few individuals ; but these are perhaps either aberrant species, or such as are on the way to becoming extinct.

"The female of *Meropogon forsteni* has not such brilliant colours as the male, and the lengthened feathers of the throat are not as handsome. But I cannot agree with Mr. Wallace's opinion, above cited, as to its nearest ally being in Africa. The species of *Nyctiornis* of the Malay archipelago are its most natural and nearest allies; and *Meropogon forsteni* gives to Celebes no other characteristic feature than *Nyctiornis amictus* gives to Borneo and Sumatra. All these are alike related to the West-African Bee-eaters, belonging to one and the same family, which occupies nearly the whole Ethiopian, Oriental, and Australian region.

"There is some error in Lord Tweeddale's (*l. c.* p. 42) giving the habitat 'Rurukan' on my authority in a paper read May 2nd, 1871, in London; whereas I only obtained my first specimen in North Celebes itself at the end of the same month."

In conclusion, I may remark that M. Oustalet, the Curator of the Museum at the Jardin des Plantes in Paris, told me that he purchased several examples of this rare bird from a plumassier, who had received them with other bright-coloured birds, and would have cut them up for plumes had not M. Oustalet fortunately rescued them from so sad a fate.

The specimen figured and described is the one in the Tweeddale collection, for the loan of which I am indebted to Captain R. G. Wardlaw Ramsay.

In the preparation of the above article I have examined the following specimens :—

E Mus. Tweeddale.

a, ♂. Rurukan, Celebes.

E Mus. Paris.

a, b. Celebes.

E Mus. Lugd.

a, ♂. Tondano, Celebes (*Dr. Forsten*).

E Mus. Acad. Dublin.

a. Celebes.

D

Genus MEROPS.

Merops, Linnæus, Syst. Nat. i. p. 182 (1766). Type *Merops apiaster*.
Tephraerops, Reichenbach, Meropinæ, p. 82 (1852). Type *Merops malimbicus*.
Melittotheres, Reichb. ut suprà (1852). Type *Merops nubicus*.
Blepharomerops, Reichb. ut suprà (1852). Type *Merops persicus*.
Phlothrus, Reichb. ut suprà (1852). Type *Merops viridis*.
Aerops, Reichb. ut suprà (1852). Type *Merops albicollis*.
Melittophas, Reichb. ut suprà (1852). Type *Merops bicolor*.
Melittophagus, Reichb. ut suprà (1852), nec Boie.
Cosmaerops, Cabanis, Mus. Hein. ii. p. 138 (1859). Type *Merops ornatus*.
Bombylonax, Heine, Journ. für Orn. 1859, p. 484 (1859). Type *Merops breweri*.
Archimerops, Hartlaub in Wiegm. Archiv, xxvi. p. 90 (1860). Type *Merops breweri*.
Meropogon, Cassin, Proc. Acad. Nat. Sc. Phil. 1859, p. 39 (1859), nec Bp.

HAB. The southern part of the Palæarctic Region, the entire Ethiopian and Indo-Malayan Regions, and the Australian Region.

Alis longis, acutis; remige primâ brevissimâ, secundâ omnium longissimâ, tertiâ breviore, scapularibus brevioribus : caudâ æquali, rectricibus duabus centralibus valde elongatis : rostro elongato, gracili, curvato : juguli plumis haud elongatis : pedibus brevibus, robustis.

Bill long, somewhat slender, curved, pointed, pentagonal at the base, then four-sided, compressed ; gape-line curved ; nostrils roundish, nasal membrane short. Wings long, pointed, the first quill very small, the second longest, the third rather shorter ; scapulars shorter than the secondaries. Tail long, even, the two central rectrices elongated and pointed. Feet small, feeble, the lower part of the tibia bare, the tarsus indistinctly scutellate ; toes short, slender, the anterior parallel and partly united ; claws slender, curved, compressed, acute.—Type *Merops apiaster*.

THE present genus contains fifteen species, which are widely distributed throughout the Old World. They are, as a rule, gregarious, and frequent open places, chiefly near water or on the borders of forests, and feed on insects, which they capture on the wing ; their flight is swift and Swallow-like, and their cry is somewhat harsh and monotonous. They breed in colonies, nesting in holes in a bank or cliff, usually near, but sometimes tolerably far away from water, and excavating their own nest-holes. Their eggs are pure white, roundish, and glossy in texture, and are deposited in a chamber at the end of the nest-hole, either on the ground or on a few straws or feathers, no regular nest being made.

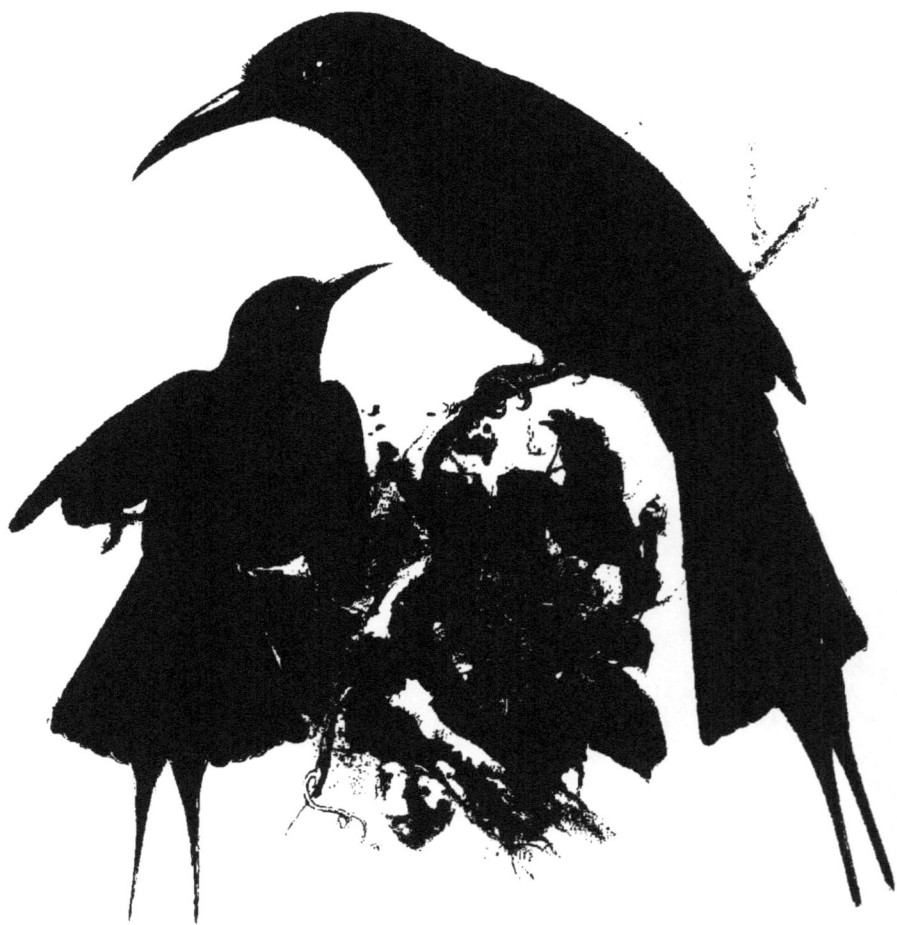

BLACK STARRY GRAFT BIRD.
NATURAL OR REST

MEROPS BREWERI.

BLACK-HEADED GREEN BEE-EATER.

Merops breweri, Cassin, Proc. Acad. Nat. Sc. Phil. 1859, p. 14; Bocage, Orn. d'Angola, p. 537 (1881).

Bombylonax breweri (Cass.), Heine, J. f. Orn. 1859, p. 434; G. R. Gray, Hand-l. of B. i. p. 98, no. 1200 (1869); Reichenow, J. f. Orn. 1877, p. 21.

Meropogon breweri, Cass. Proc. Acad. Nat. Sc. Phil. 1859, p. 39; Heine, J. f. O. 1859, p. 433; Cass. J. Acad. N. Sc. Phil. 2nd ser. iv. p. 321, pl. 49. fig. 1 (1860); Sharpe & Bouvier, Bull. Soc. Zool. France, i. p. 40 (1876).

Archimerops breweri (Cass.), Hartl. in Wiegm. Archiv, xxvi. p. 90 (1860).

Figura unica.

Cassin, J. Acad. Nat. Sc. Phil. 2nd ser. iv. pl. 49. fig. 1.

Had. West Africa.

Ad. capite et collo saturatè nigris: corpore et alis suprà cum rectricibus centralibus duabus elongatis, saturatà psittacino-viridibus: caudâ castaneo-rufâ, rectricibus externis utrinque in pogonio externo saturatè viridibus: corpore subtùs aurantiaco vix viridi tincto: rostro et pedibus nigris: iride rubrâ.

Adult male (Landana, W. Africa).—Head and neck deep black; upper parts generally and the two central elongated rectrices rich deep parrot-green; tail deep fox-red tipped with green, and the outer web of the external feather on each side deep green; underparts deep golden orange with a greenish wash; beak and legs black, iris red. Total length about 12 to 13 inches, culmen 1·9, height of culmen at base 0·4, wing 4·6, tail 6·4, central rectrices extending 1·9 beyond the lateral ones, tarsus 0·55.

Adult female.—Does not differ from the male.

BUT little is known respecting this rare species, and it is only lately that specimens have been obtainable through dealers. It was first obtained by DuChaillu on the Gaboon, and was described by Cassin from a specimen brought back by him. It is stated by DuChaillu to occur on the Camma and Ogobai rivers in Gaboon; Dr. Reichenow records it from the Loango coast; and Professor Barboza du Bocage writes (*l. c.*) that it inhabits the coast of Loango and Chinchouxo, giving as his authorities Dr. Falkenstein and M. L. Petit aîné. This latter gentleman met with it on the Congo; and I am indebted to him for the following notes respecting its habits:—"This pretty black-headed Bee-eater is not common, and on only one occasion did I meet with about forty individuals perched on a tree on the banks of the Chiloango river near Landana, and was fortunate enough to kill two, the rest escaping; they were evidently on passage. I usually met with this species in the depths of the forests in open places, and near the water alongside the patches cultivated by the natives. It is an easy bird to shoot, and when I met a male and female, I usually succeeded in obtaining both. It darts with rapidity amongst the bushes, where it can fly with ease, and catches the beautiful shining insects (which I have always found when examining the contents of the stomachs of these birds) ; and I have often seen it perched on a species of tree which has a bunch of leaves at the end of each branch, which forms a fan. From time to time it utters a cry like that of the little banded squirrel."

This species has been separated generically by several authors from the true Meropidæ; but, so far as I can judge, the differences are not sufficient to warrant such separation. These differences I may give as follows :—Beak strong, somewhat higher than in *Merops apiaster*, but not so high as in *Nyctiornis*, slightly more curved than in *Merops*, not grooved. Wings short and broad, not reaching far beyond the base of the tail; first quill short, fully 1·8 inch shorter than the second, which is 0·42 shorter than the third; the third, fourth, and fifth nearly equal, the fourth being longest. Tail slightly rounded, the two central rectrices being, however, much elongated and attenuated. Tarsus as in *Merops apiaster*. Altogether this species differs from the rest of those included in the genus *Merops* in having the wings shorter and broader, the other differences being too slight to take into consideration.

The specimen above described and figured is one in my own collection from Landana, West Africa.

In the preparation of the above article I have examined the following specimens :—

E Mus. H. E. Dresser.

a, ♂ ad. Landana, Congo.

E Mus. Brit.

a, ad. Gaboon.

E Mus. G. E. Shelley.

a. Congo (*Dr. A. Lucan*).

SUMATRAN BEE EATER
MEROPS SUMATRANUS

MEROPS SUMATRANUS.

SUMATRAN BEE-EATER.

Merops sumatranus, Raffles, Trans. Linn. Soc. xiii. p. 294 (1821); Steph. in Shaw's Gen. Zool. xiii. part 2, p. 73 (1825); Gray, Gen. of B. i. p. 86 (1846); Blyth, Cat. Mus. As. Soc. p. 52 (1849); Gould, P. Z. S. 1859, p. 151; Tweeddale, Trans. Zool. Soc. ix. p. 151, pl. xxvi. fig. 2 (1875); id. Ibis, 1877, p. 297; Sharpe, Ibis, 1879, p. 248; Hume, Stray Feathers, viii. p. 48 (1879); Tiraut, Bull. Soc. Comm. Agric. de la Cochin Chine, sér. 3, vol. i. p. 98 (1879).

Merops cyanopygius, Less. Traité d'Orn. p. 238 (1831).

Le Guêpier de Sumatra, Less. Compl. Buff. ii. Ois. 2nd ed. pl. fig. 2 (1840).

Merops bicolor, Horsfield & Moore, Cat. B. E.I. Co. Mus. i. p. 87 (1854), nec Bodd.; Mottley & Dillwyn, Contrib. Nat. Hist. Labuan and Borneo, p. 14 (1855); Salvadori, Ucc. Born. p. 90 (1874); Sharpe, P. Z. S. 1875, p. 101; id. Ibis, 1877, p. 5; David & Oust. Ois. de la Chine, p. 73 (1877); Kelham, Ibis, 1881, p. 377.

Merops badius, Bp. Consp. Gen. Av. i. p. 162 (1850), nec Gmel.; Licht. Nomencl. Av. p. 66 (1854); Schlegel, Mus. Pays-Bas, *Merops*, p. 3 (1863, partim); Sclater, P. Z. S. 1863, p. 213; Pelzeln, Novara Reise, Vög. i. p. 50 (1865); Kelham, Ibis, 1881, p. 377.

Melittophas bicolor, Cab. Mus. Hein. ii. p. 157 (1859), nec Bodd.; Gray, Hand-l. of B. i. p. 99, no. 1202 (1869, partim).

? *Melittophas chlorolæmus*, Gray, Hand-l. of B. i. p. 99, no. 1203 (1869).

Merops rochechouardi, Heude, Ann. Sc. Nat. 1873 [from label on specimen in Paris Museum].

Figura unica.

Tweeddale, Trans. Zool. Soc. ix. pl. xxvi. fig. 2.

Hab. China, Cochin China, Malay peninsula, Java, Sumatra, and Borneo.

Ad. *Meropi bicolori* persimilis, sed capite et dorso valdè saturatioribus : mento et gulâ saturatè cæruleis nec viridi tinctis.

Juv. capite, nuchâ et dorso saturatè viridibus nec castaneis : mento et gulâ sordidè cæruleis vix viridi tinctis : abdomine imo albo pallidè viridi tincto : subcaudalibus cæruleo tinctis, rectricibus centralibus non elongatis.

Adult male (Qualla Kangsa, Perak, 25th February).—Resembles *Merops bicolor*, but has the head, neck, and back much darker, the colour being deep rich chestnut-brown, with an almost coppery tinge; upper throat deep blue, without any trace of green, the blue being sharply divided from the green on the rest of the underparts; bill, legs, and iris as in *Merops bicolor*. Total length about 9½ inches, culmen 1·55, wing 4·4, tail 5·3, central rectrices extending 2·1 inches beyond the lateral ones, tarsus 0·5.

Adult female (Cebu).—Does not differ from the male in coloration of plumage.

Young (Lampong).—Crown, nape, and back deep green, not brown; upper throat dull blue, slightly tinged with green; lower abdomen white, tinged with blue; central tail-feathers not elongated; rest of the plumage as in the adult, but duller.

Obs. A specimen in the Tweeddale collection which appears to me to be in change from the immature to the fully adult dress has the chestnut-brown intermixed with dark green, the green on the wings marked with deep blue, and the green underparts blotched with blue; the central rectrices are elongated, but much abraded, as, indeed, is much of the rest of the plumage.

Nestling (Labuan).—Resembles the young bird above described, but has the throat of a much deeper blue colour, without any trace of green, and is, indeed, much more deeply coloured than the immature example figured and above described. This nestling bird, for which I am indebted to Mr. R. Bowdler Sharpe of the British Museum, was received too late to be figured, which is the more to be regretted as it is a most characteristic specimen, showing, as it does, the blue on the throat so much developed in the nestling plumage.

This Bee-eater has been so generally confused with *Merops bicolor* that it is rather difficult to define its range; but, so far as I can judge, it inhabits China, Cochin China, Siam, the southern Malay peninsula, Sumatra, Borneo, and probably also Java.

In the Paris Museum there is a specimen, which is clearly referable to this species, which was obtained by Père Heude at Kiangsi, in China, in June 1872, and which is marked as being the type of *Merops rochechouardi* of Heude. M. Tiraut states (*l. c.*) that he killed one at Trà-vinh in Cochin China, which country it is said to inhabit generally, but it does not appear to be common there. Mr. Hume (Str. Feath. viii. p. 48) records it as occurring at Malacca, Pulo Sehan, Kurroo, Chohong, and Singapore; and Lieut. Kelham (*l. c.*) met with it on the banks of the Perak river, and also at Malacca and Singapore. I have examined specimens from Malacca, Sumatra, and Borneo, and can fully indorse the statement by Lord Tweeddale that examples from these localities do not differ. Governor Ussher obtained it at Lumbidan in Borneo; and Mr. Everett speaks of it (Ibis, 1877, p. 5) as being a common bird in that island. On the strength of a specimen in the Paris Museum which is said to have been obtained in Java, it is recorded as found in that island: Lord Tweeddale, however, doubts its occurrence there; but it appears to me that there is every probability that it does inhabit that island.

In the Philippine Islands it is replaced by its near ally *Merops bicolor*, and, with one exception, I can find no instance of the occurrence of *Merops sumatranus* in that group of islands. When examining the Bee-eaters in the Paris Museum, I found one specimen labelled as having been obtained by MM. Hombron and Jacquinot at Jolajola, in the island of Luçon, which is undoubtedly referable to *M. sumatranus* and not to *M. bicolor*. This is the more puzzling, because all the other specimens I have seen from the Philippines are referable to *Merops bicolor*, and all the information I have gathered (with this one exception) tends to show that *M. bicolor* alone is found there; thus I cannot help suspecting that there is a mistake as regards the locality where this specimen was really procured.

Writing on the habits of this bird as observed by him in Borneo, Mr. Mottley says that " these birds come to Labuan to breed, which they do in deep holes dug in the sand; they all leave when the ruins begin. They principally haunt those places where there is a small open grassy spot on the sea-shore, associating in flocks of ten or twelve, and are extremely shy and difficult to approach ; they sail in circles with the flight of a small hawk, sometimes at a great height, and sometimes close to the grass ; when they perch, which is not often, they usually select a bare twig. I kept a young one alive for some time, and fed him on cockroaches and grasshoppers, and he became exceedingly tame ; he was, however, at last killed by eating a large spider, which evidently poisoned him." Mr. Mottley adds that he found it uncommon in Borneo. Mr. Everett, however, says (Ibis, 1877, p. 5) that in Borneo it is "an abundant species, but confined to the sandy tracts on the shore-line, though a pair will be met with now and again as far as 20 miles inland, where a sandy bank happens to offer facilities for nidification . . . . The flight of these birds is strong, and combines the swift skimming of the Swallow with the airy hovering of the Falcon. Now they will flutter up just as a Skylark does, and then swoop earthwards like a Hawk after its quarry, and then again will rise and float almost without motion, merely balancing themselves in the breeze by a slight quivering of the pinions. When at rest they commonly perch on the topmost twigs of the lower *Casuarina* trees. The gizzard always contains insects—beetles, dragon-flies, and orthoptera, as well as wasps and bees."

Respecting the nidification of this Bee-eater I find nothing on record ; but it doubtless, like its congeners, makes its nest in a hole tunnelled in a bank, and deposits white eggs.

The specimens figured are the adult and young birds above described, the former of which is in my own collection, and the latter in the Tweeddale collection.

In the preparation of the above article I have examined the following specimens :—

E Mus. H. E. Dresser.

a, ♂ ad. Qualla Kangsa, Perak, 25th February, 1877 (*Lieut. Kelham*). *b.* Qualla Kangsa, 20th June, 1877 (*Kelham*).

E Mus. Tweeddale.

a, b, ad. ; *c,* juv. ; *d,* juv. Malacca (*Wardlaw Ramsay*). *e,* ad. N.E. Borneo. *f, g, h, i.* Lampong, S.E. Sumatra. *k, ♂.* Sumatra, 16th September, 1878 (*Bock*). *l,* juv. Sumatra (*Bock*). *m,* juv. Lampong.

E Mus. Paris.

a, ♀. Kiangsi, June 1872 (type of *M. rochechouardi*). *b.* Java (*Diard*). *c.* Sumatra (*Duvaucel*) : type of *M. cyano-pygius*, Less. *d,* juv. Sumatra (*Duvaucel*).

E Mus. Brit.

a, b. Malacca (*Evans*). *c.* Penang, April 1854 (*Dr. Cantor*). *d.* Sumatra, 1861 (*Wallace*). *e.* Sumatra (*Gould coll.*). *f, g.* Labuan (*Low*). *h, ♀.* Sarawak, 21st May, 1870 (*Everett*). *i.* Borneo (*Mottley*).

CHESNUT AND GREEN BEE EATER

M. ACUT BICOL ER

MEROPS BICOLOR.

CHESTNUT-AND-GREEN BEE-EATER.

Apiaster ex Franciæ insula, Briss. Oru. iv. p. 543, pl. xliv. fig. 2 (1760).

Apiaster philippensis minor, Briss. tom. cit. p. 555, pl. xliii. fig. 2 (1760).

Merops americanus, P. L. S. Müller, Natursyst. Suppl. p. 95 (1776).

Le Guêpier marron et bleu, Montb. Hist. Nat. Ois. vi. p. 493 (1779).

Guêpier de l'isle de France, D'Aubenton, Pl. Enl. 252.

Merops bicolor, Boddaert, Tabl. des Pl. Enl. p. 15, no. 252 (1783, ex d'Aubenton) ; Tweeddale, Trans. Zool. Soc. ix. p. 150 (1875) ; id. P. Z. S. 1877, pp. 540, 690, 757, 822 ; id. P. Z. S. 1878, pp. 282, 340.

Merops badius, Gmel. Syst. Nat. i. p. 462, no. 10 (1788, ex Briss.) ; Gray, Gen. of B. i. p. 86 (1846) ; F. Moore, P. Z. S. 1854, p. 264 ; Schlegel, Mus. Pays-Bas, *Merops*, p. 3 (1863, partim).

Merops castaneus, Lath. Ind. Orn. i. p. 273, no. 10 (1790) ; Vieill. Nouv. Dict. xiv. p. 18 (1816).

Le Guêpier marron et bleu ou le Guêpier Latreille, Levaill. Hist. Nat. Ois. p. 45, pl. 12 (1807).

Merops erythrocephalus, Vieill. Nouv. Dict. xiv. p. 24 (1816).

Merops hypoglaucus, Reichenb. Meropinæ, p. 76 (1852).

Melittophas badius (Gm.), Reichenb. Meropinæ, p. 82 (1852) ; Cab. Mus. Hein. ii. p. 137 (1859–60, partim) ; Gray, Hand-l. of B. i. p. 99, no. 1202 (1869, partim).

Melittophus hypoglaucus, Reichenb. Meropinæ, p. 82 (1852).

Merops ornatus, v. Martens, J. f. Orn. 1866, p. 17, nec Lath.

Figuræ notabiles.

D'Aubenton, Pl. Enl. 252 ; Levaillant, Hist. Nat. Ois. pl. 12 ; Reichenbach, Meropinæ, pl. ccccxlix. figs. 3243, 3244 ; Tweeddale, Trans. Zool. Soc. ix. pl. xxvi. fig. 1.

Hab. Philippine Islands.

Ad. pileo, nuchâ et dorso ferrugineo-castaneis : dorso imo supracaudalibus et subcaudalibus pallidè cærulcis : caudâ saturatè cœruleâ, rectricibus centralibus elongatis, versus apicem attenuatis et splendidiùs coloratis : alis, scapularibus et dorso centraliter saturatè psittacino-viridibus, remigibus nigricante apicatis : corpore subtùs pallidè viridi : striâ per oculos ductâ et regione paroticâ nigris, illâ suprâ turcino marginatâ : iride rubrâ : rostro nigricante : pedibus griseo-fuscis.

Juv. suprà saturatè psittacino-viridis, supracaudalibus et uropygio imo pallidè cærulcis : rectricibus æqualibus nec elongatis : subtùs pallidè viridis, vix cæruleo tinctus.

Adult male (Manilla).—Crown, nape, and fore part of the back rich fox-red ; lower back and upper tail-coverts pale blue ; tail dark blue, the central rectrices clearer and richer in colour, elongated and attenuated towards the end ; wings, scapulars, and centre of back deep parrot-green, most of the quills tipped with blackish; underparts generally pale green; under tail-coverts pale blue; a broad black streak passes from the base of the bill through the eye and covers the ear-coverts, below which is a broad turquoise streak on each side of the throat; iris rich red; bill blackish; legs purplish brown. Total length about 9 to 9½ inches, culmen 1·7, wing 4·6, tail 5·8, central rectrices extending 2·4 beyond the lateral ones, tarsus 0·5.

Adult female (Batuan).—Does not differ from the male in size or in coloration of plumage.

Young (Manilla).—Upper parts generally, except the rump and upper tail-coverts, deep dull parrot-green, the crown being rather darker than the rest of the upper parts; lower rump and upper tail-coverts pale blue; tail even, the central rectrices not elongated; underparts pale green with a bluish tinge, the under tail-coverts bluish white.

THE range of this Bee-eater appears, so far as I can judge, to be restricted solely to the Philippines, where it is tolerably numerous. It is true that it has, by several authors, been stated to occur in other localities; but, so far as I can ascertain, these statements refer to its near ally *Merops sumatranus*, and not to the true *Merops bicolor*. It has been recorded by various collectors from different parts of the Philippines as occurring from the month of February to October; but it is probably also to be met with at other seasons of the year, and is therefore resident in that group of islands. Lord Tweeddale's collectors obtained it in many localities, as will be seen by the list of specimens in his collection at the close of this article; and Dr. Steere met with it in Panay and at Dumalon, Mindanao. This latter explorer states (Trans. Linn. Soc. 2nd ser. Zool. i. p. 316) that at Dumalon he found it in swamps, perching in low trees; but in Panay it was shot more in the open, settling on weeds and sticks a little way from the ground.

I have never seen the eggs of this Bee-eater, and although by no means common in the Philippines, I do not find that any notes respecting its breeding-habits have been published; but it is fair to infer that it nests in holes in the river-banks and deposits white eggs like its allies.

The specific name for the present species, which, strictly speaking, has the priority, is that of *americanus*, given to it by P. L. S. Müller (*l. c.*), who merely gave that name to the bird figured by D'Aubenton under the name of *Guêpier de l'Isle de France*, and described by Montbeillard in Buffon's Hist. Nat. Ois. vi. p. 493. The description given by Müller is as follows, viz. :— "8. Der Blaubauch. *Merops americanus.* Der Rücken ist braun, der Bauch blau, die Flügel sind seegrün, und der Schwanz hat zwei sehr lange Ruderfedern. Der Aufenthalt ist in Isle de France. Buffon." Müller does not give any reason for calling a bird which has never been met with in the New World *americanus*; and as it is so evidently a misnomer, and has besides never been applied to this species by any other author, the best plan, it seems to me, will be to let it remain as a mere synonym.

The specimens figured are the adult and young birds above described, which are both in my own collection.

In the preparation of the above article I have examined the following specimens:—

E Mus. H. E. Dresser.

a, ♂ ad. ; *b*, juv. Manilla (*Boucard*).   *c*, ♂. Cebu, Philippines, April 1877 (*Everett*).   *d*. Philippines, October 1874 (*Gerrard*).

E Mus. Brit.

a, b. Philippines (*Gould coll.*).   *c*. Panay (*Steere*).

E Mus. Tweeddale.

a, ♂. Valencia, August 1877.   *b, c, d, e*. Valencia.   *f*, ♂ ; *g*, ♀. Cebu, March and April 1877 (*Everett*).   *h*, ♂ ; *i, k*, ♀. Luçon, April (*Meyer*).   *l, m*, ♂. S. Leyte, September 1877.   *n*, ♀. Batuan, May 1877.   *o*, ♂. Placer, July 1877.   *p*. Monte Alban, March 1877.

E Mus. Paris.

a. Luçon (*Verreaux*).   *b*. Manilla, 1844 (*Voyage of 'Favorite'*).

LITTLE GREEN BEE EATER
MEROPS VIRIDIS
(EGYPT)

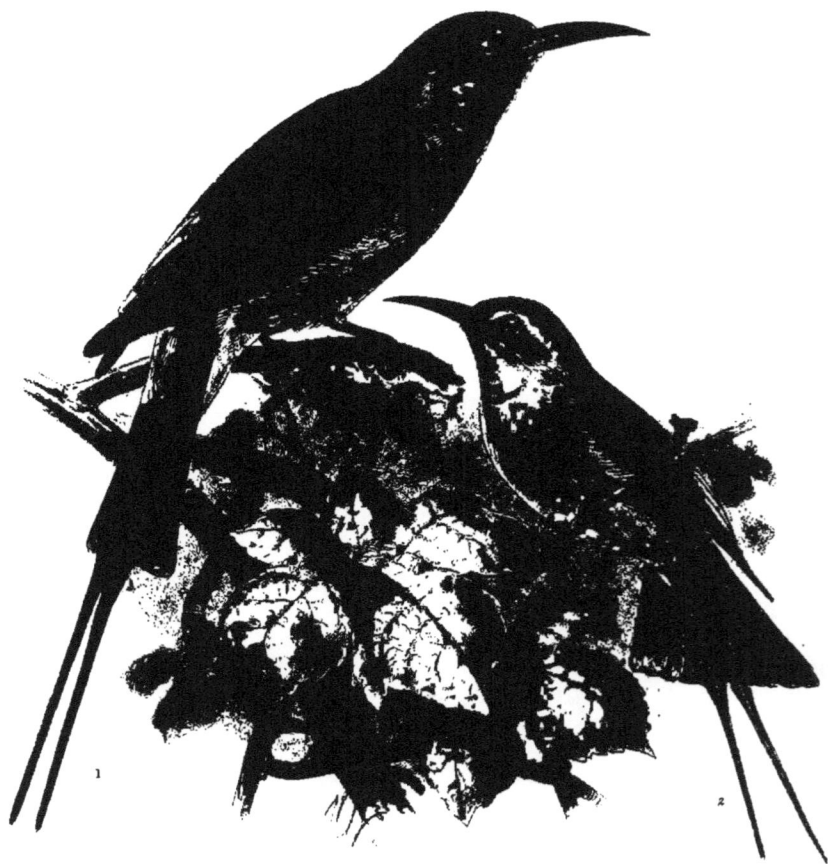

LITTLE GREEN BEE EATER
MEROPS VIRIDIS
(1 CEYLON 2 BURMAH)

MEROPS VIRIDIS.

LITTLE GREEN BEE-EATER.

Apiaster madagascariensis torquatus, Brisson, Orn. iv. p. 549 (1760).

Apiaster bengalensis torquatus, Brisson, tom. cit. p. 552 (1760).

Merops viridis, Linn. Syst. Nat. i. p. 282 (1766) ; Gmel. Syst. Nat. i. p. 461 (1788) ; Lath. Ind. Orn. p. 270 (1790) ; Bonnaterre, Tabl. Encycl. et Méthod. p. 273 (1790); Licht. Cat. rer. nat. rariss. p. 21 (1793) ; Shaw, Gen. Zool. viii. pt. 1, p. 156 (1811); Vieill. Nuov. Dict. xiv. pp. 14 & 26 (1816) ; Steph. in Shaw's Gen. Zool. xiii. pt. 2, p. 73 (1825); Lesson, Traité d'Orn. p. 238 (1831) ; Franklin, P. Z. S. 1831, p. 115 ; Sykes, P. Z. S. 1832, p. 82 ; Ewer, P. Z. S. 1842, p. 92; Strickland, P. Z. S. 1842, p. 167 ; Rüpp. Syst. Ueber. p. 24 (1845) ; Gray, Gen. of B. i. p. 86 (1846); Blyth, Cat. B. Mus. As. Soc. p. 53 (1849); Bp. Consp. Gen. Av. i. p. 162 (1850); Kelaart, Prodromus, Cat. p. 119 (1852); Licht. Nomencl. Av. p. 66 (1854) ; Horsf. & Moore, Cat. B. E.I. Co. Mus. i. p. 84 (1854) ; Gould, B. of Asia, pt. 7 (1855); Moore, P. Z. S. 1857, p. 87 ; A. L. Adams, P. Z. S. 1858, p. 475 ; id. P. Z. S. 1859, p. 174 ; E. C. Taylor, Ibis, 1859, p. 47; Gould, P. Z. S. 1859, pp. 150, 151 ; Irby, Ibis, 1861, p. 228 ; Jerdon, B. of India, i. p. 205 (1862) ; Tristram, Ibis, 1862, p. 278 ; Schlegel, Mus. Pays-Bas, *Merops*, p. 6 (1863) ; Tristram, P. Z. S. 1864, p. 433; Swinhoe, Ibis, 1864, p. 415 ; A. L. Adams, Ibis, 1864, p. 15 ; Beavan, Ibis, 1865, p. 407; Walden, P. Z. S. 1866, p. 538 ; E. C. Taylor, Ibis, 1867, p. 55 ; Pelzeln, Ibis, 1868, p. 307 ; Beavan, Ibis, 1869, p. 407 ; Blanford, Ibis, 1870, p. 465; Blyth, Ibis, 1870, p. 162; Blauf. Geol. & Zool. of Abyss. p. 320 (1870) ; Shelley, B. of Egypt, p. 171 (1872) ; Jerdon, Ibis, 1872, p. 2 ; Holdsworth, P. Z. S. 1872, p. 422 ; Hume, Nests & Eggs of Ind. B. p. 99 (1873) ; Hume, Stray Feathers, i. p. 167 (1873) ; Adam, tom. cit. p. 371 (1873) ; Hayes Lloyd, Ibis, 1873, p. 406; Ball, Str. Feath. ii. p. 386 (1874) ; Hume, tom. cit. p. 469 (1874) ; Hume, Str. Feath. iii. p. 455 (1875); Oates, tom. cit. p. 49 (1875) ; Legge, Ibis, 1875, p. 281 ; R. W. Morgan, Ibis, 1875, p. 314 ; Fairbank, Str. Feath. iv. p. 254 (1876) ; Armstrong, Str. Feath. iv. p. 304 (1876) ; Dresser, B. of Europe, v. p. 171 (1876) ; Blanf. E. Pers. ii. p. 124 (1876) ; Hume, Str. Feath. v. p. 18 (1877) ; Fairbank, tom. cit. p. 394 (1877) ; Hume, Str. Feath. vi. p. 498 (1878) ; id. op. cit. vii. pp. 35, 77 (1878) ; Ball, tom. cit. p. 203 (1878) ; Cripps, tom. cit. p. 258 (1878) ; Scully, tom. cit. p. 237 (1878) ; Anderson, Yunnan Exped. i. p. 582 (1878) ; Tiraut, Bull. Soc. Com. Agric. de la Cochin Chine, sér. 3, i. p. 97 (1879) ; Legge, B. of Ceylon, p. 309 (1880) ; Bingham, Str. Feath. ix. p. 152 (1880) ; Vidal, tom. cit. p. 48 (1880) ; Oates, B. of Brit. Burmah, ii. p. 65 (1883).

Le Guêpier vert à gorge bleu, Montb. Hist. Nat. Ois. vi. p. 497 ; Pl. Enl. 259 (1779).

Le Guêpier jaune de Coromandel, Sonnerat, Voy. Ind. ii. p. 213, pl. 119 (1782).

Merops coromandus, Lath. Ind. Orn. i. p. 272 (1790).

? *Merops flavicans*, Lath. Ind. Orn. p. 272 (1790) ; Bonnaterre, Tabl. Encycl. et Méthod. p. 277 (1790).

Merops orientalis, Lath. Ind. Orn. Suppl. p. 32 (1801) ; Shaw, Gen. Zool. viii. pt. 1, p. 178 (1811) ; Gray, Gen. of B. i. p. 86 (1846).

Merops cærulescens, Lath. Ind. Orn. Suppl. p. 33 (1801).

Le Guêpier à gorge bleu mâle ou le Guêpier Lamarck, Levaill. Hist. Nat. Guêp. p. 39, pl. 10 (1807).

Le Guêpier citrin, Levaill. op. cit. p. 41, pl. 11 (1807).

Merops lamark, Cuvier, Règne Animal, i. p. 442 (1829).

Merops viridissimus, Swainson, B. of W. Afr. ii. p. 82 (1837) ; Bp. Consp. Gen. Av. i. p. 162 (1850) ; von Müller,

J. f. Orn. 1855, p. 10; Hartlaub, Orn. Westafr. p. 40 (1857); id. J. f. Orn. 1860, p. 87; id. Faun. Madag. p. 32 (1861); Heuglin, J. f. O. 1860, p. 334; Schlegel, Mus. Pays-Bas, *Merops*, p. 6 (1863); Hartmann, J. f. O. 1866, p. 202; Heuglin, Orn. N.O.-Afr. i. p. 202 (1869); Gray, Hand-l. of B. i. p. 100, no. 1211 (1869); Finsch, Tr. Zool. Soc. vii. p. 224 (1870); Shelley, Ibis, 1871, p. 48.

Merops indicus, Jerdon, Madr. Journ. ser. 1, xi. p. 227 (1840); Burgess, P. Z. S. 1855, p. 27.

Merops ferrugeiceps, Hodgs. in Gray's Zool. Misc. p. 82 (1844).

Merops torquatus, Hodgs. in Gray's Zool. Misc. p. 84 (1844).

Merops luteus, Reichenbach, Meropinæ, p. 75 (1852).

Phlothrus viridissimus (Swains.), Reicheub. Meropinæ, p. 82 (1852); Cab. Mus. Hein. ii. p. 137 (1859); Gray, Hand-l. of B. i. p. 100, no. 1211 (1869).

Phlothrus viridis (Linn.), Reicheub. Meropinæ, p. 82 (1852); Cab. Mus. Hein. ii. p. 136 (1859); Gray, Hand-l. of B. i. p. 99, no. 1210 (1869).

Merops lamarcki (Cuv.), Strickland, P. Z. S. 1850, p. 216; Sclater, in Jard. Contrib. Orn. ii. p. 124 (1854).

Hurrial, Patringa, Hind.; *Bansputtee*, lit. "Bamboo-leaf," Bengal.; *Chinna passeriki*, Tel., lit. "Small green bird" (Jerdon); *Mo-na-gyee*, Arracan (Blyth); *Kurumenne kurulla*, Sinhalese; *Kattalan kuruvi*, Tamils in Ceylon (Legge).

Figuræ notabiles.

D'Aubenton, Pl. Enl. 259; Sonnerat, Voy. Ind. ii. pl. 119; Levaillant, Hist. Nat. Guêp. pls. 10, 11; Gould, B. of Asia, pt. vii.; Dresser, B. of Europe, v. pl. 297; Shelley, B. of Egypt, pl. vii. fig. 2.

HAB. N. and N.E. Africa, Palestine, India, eastward to Cochin China.

Ad. suprà lætè psittacino-viridis, pileo et nuchâ vix rufescente aureo tinctis: dorso imo, uropygio et scapularibus cum secundariis intimis viridi-cyaneo lavatis: remigibus intùs viridi-cinnamomeis, primariis nigro-fumoso et secundariis nigro terminatis, secundariis intimis dorso concoloribus: caudâ sordidè viridi, plumis in pogonio interno cinnamomeo tinctis, rectricibus duabus centralibus valdè elongatis et angustatis: corpore subtùs lætè psittacino-viridi: tæniâ transoculari et fasciâ jugulari angustâ nigris: gulæ lateribus cæruleis, et gulâ centraliter vix cæruleo notatâ: abdomine cæruleo et viridi-cæruleo tincto: alis infrà pallidè fulvescentibus: rostro nigro: iride scarlatinâ: pedibus sordidè plumbeis.

Juv. vix sordidior et pallidior: gulâ magis viridi-flavâ et fasciâ jugulari angustiore.

Adult male (Egypt).—Upper parts brilliant green; crown and nape tinged with rufescent golden; lower back, rump, scapulars, and inner secondaries tinged with verdigris or blue-green; quills rufous on the inner webs, and strongly tinged with green on the outer webs, the primaries dusky at the tips, and the secondaries, except the innermost, broadly tipped with black; innermost secondaries like the back; tail dull green, the inner edge of the webs dusky rufescent, the two central feathers much elongated; underparts bright green; a black band through the eye, and a collar on the lower throat black; a line below the eye-band bright blue, and the throat and abdomen slightly tinged with blue and blue-green; under surface of the wings bright rufous; bill blackish; iris crimson; legs dull plumbeous. Total length about 11 inches, culmen 0·95, wing 3·65, lateral tail-feathers 2·95, central tail-feathers 5·35, tarsus 0·45.

Adult female (Egypt).—Resembles the male; but the throat and underparts are green, and the line of blue below the black eye-band is narrow.

Young.—Rather duller and paler than the adult, the throat pale yellowish green, and the collar on the throat smaller and less distinct; the blue stripes on each side of the neck are wanting.

Adult male (Ceylon).—Differs from the male from Egypt, above described, in having the crown, nape, and hind neck deep golden rufous and the throat verditer-blue.

Adult male (Burmah).—Differs from the ordinary type in having the forehead, crown, nape, upper back, and ear-coverts chestnut, tinged with green on the forehead and crown; beneath the black eye-streak a line of bright blue; cheeks, chin, and throat greenish blue.

This brilliantly coloured little Bee-eater inhabits the whole of the northern and eastern portions of the Ethiopian Region, ranging just into the Western Palæarctic Region, and the Oriental Region as far east as Cochin China and as far south as Ceylon. I included it in my work on the Birds of the Western Palæarctic Region on the strength of its occurrence in Palestine and North-east Africa. Canon Tristram ('Ibis,' 1862, p. 278) obtained it in the Jordan valley. It is a common North-east African species and ranges there northward into the limits of the Western Palæarctic Region. Mr. J. H. Gurney, jun., informs me that when travelling up the Nile he met with it two days after leaving Cairo, whence it was common up to Assouan; and Captain Shelley states (B. of Egypt, p. 171) that it is "a resident in Middle Egypt throughout the year, but does not during the winter months range north of Golosaneh. They do not congregate in flocks, like *Merops apiaster* and *Merops ægyptius*, but are generally to be met with in pairs or family parties, often perched in rows on the long leaves of the date-palms, or on the outer twigs of the sont trees. In flight they look extremely beautiful, as they skim gracefully through the air with outspread wings, showing the orange colour underneath like an illuminated transparency. They breed in holes in the banks in April."

Von Heuglin writes (Orn. N.O.-Afr. i. p. 203) that this Bee-eater is a resident in Central Egypt between 24° and 28° N. lat., and not uncommon along the Nile and its canals, where it breeds from February to April. In the autumn he observed it in Eastern Kordofan, in the spring along the Gazelle river, but he cannot say if it is sedentary there; on the other hand, he believes that it breeds in the Bogos country, as it was observed there during the whole of the rainy season. Von Hartmann observes (*l. c.*) that Antinori shot several specimens at Djergele, on the Nile, in 26° 25' N. lat., and he himself obtained several in full breeding-dress at Antub, on the Blue Nile, in September 1859. Mr. Blanford found it common in the coast regions of Abyssinia, more especially in the mangroves on the shores of Annesley Bay; and other naturalists who have visited that country speak of it as being common there. It has been stated to occur in Madagascar; but, as pointed out by Dr. Hartlaub, Brisson is the only authority on this point; and as it has never been sent by any of the collectors who have more recently explored that island, it may be doubted whether it is really to be met with there. It occurs rarely in North-west Africa. There is a specimen in the Heine collection which was obtained in Barbary. Verreaux records it from Senegambia; the specimen on which Swainson's specific title *viridissimus* was founded is said to have been sent from Senegal; and it has been obtained in the Gaboon. There is, I may remark, a specimen in the Gould collection, now in the British Museum, said to have been obtained in Mauritius. In Asia it is found as far east as Cochin China. Mr. Blanford states (*l. c.*) that in Persia it is, of course, non-migratory, and is only found in the lowlands of Southern Persia and Baluchistan, to which Sir Oliver St. John adds that he found it common on the coast and in the neighbouring valleys up to 2000 feet. According to

F

Dr. Jerdon (B. of India, i. p. 205) the Little Green Bee-eater " is found over the whole of India, extending to Arrakan, the Indo-Chinese countries, and to Ceylon. It does not ascend mountains, to any height at least; and the specimen in the British Museum from Mr. Hodgson, marked from Darjeeling, assuredly never was killed there, though it occurs at the foot of the hills.

"It is a very common bird, and is a most characteristic adjunct of Indian scenery. It generally hunts, like the Flycatchers, from a fixed station, which may be the top branch of a high tree or a shrub, or hedge, a bare pole, a stalk of grain or grass, some old building, very commonly the telegraph-wires, or even a mound of earth on the plain. Here it sits, looking eagerly around, and on spying an insect, which it can do a long way off, starts rapidly, and captures it on the wing with a distinctly audible snap of its bill; it then returns to its perch, generally slowly sailing with outspread wings, the copper burnishing of its head and wings shining conspicuously, like gold, in the sun's beams. Sometimes it may be seen alone, or in small parties seated near each other, but hunting quite independently. It frequently takes two or three insects before it reseats itself on its perch; and in the morning and evening they collect in considerable numbers, and, often in company with Swallows, hawk actively about for some time. I have seen one occasionally pick an insect off a branch, or a stalk of grain or grass; and Mr. Blyth informs me that he had seen a number of them assembled round a small tank, seizing objects from the water in the manner of Kingfishers.

"They roost generally in some special spot, sometimes a few together in one tree; but at some stations all the birds for some miles round appear to congregate and roost in one favoured locality. The bamboo tope at Saugor is a celebrated spot of this kind: here Crows, Mynas, Parrakeets, Bee-eaters, Sparrows, &c. collect from miles around; and the noise they make towards sunset and early in the morning is deafening.

"The Bee-eater has a loud, rather pleasant, rolling, whistling note, which it often repeats, especially in the morning and towards evening, and often whilst hunting. They sometimes collect in small parties towards sunset on a road, and roll themselves about in the sand and dust, evidently with great pleasure.

"They breed in holes, in banks of ravines or of rivers, and on roadsides, laying two to four white eggs. Burgess mentions that in a nest that he examined there were three young ones all of different ages. They breed from March to July according to the locality, earlier in the north of India, later in the south. Mr. Blyth observed them breeding near Moulmein as late as the middle of August."

I find this species recorded from many parts of India. Captain Butler says that it occurs in abundance all over the plains of Northern Guzerat, and he also met with it in considerable numbers at Aboo, but does not think that it remains on the hills in the hot weather. In Oudh and Kumaon it is, according to Col. Irby, " excessively numerous throughout the year: ten or more may often be seen sitting on the same bush; and on the telegraph-wires on the Grand Trunk Road I once saw, in the early morning, upwards of fifty within twenty yards. In one habit this bird resembles our Spotted Flycatcher (*Muscicapa grisola*): it is incessantly flying a few feet in chase of insects, and settling again on its former perch." Captain Beavan states (*l. c.*) that it is " common at Barrackpore in the cold weather, arriving about the end of October. Breeds in Maunbhoom, where it is tolerably common, at the beginning of April. The eggs, two in number, are very round and of a pure clear white. The nest-hole is excavated in the ground." According to Burgess (*l. c.*) it is " a common bird in the Deccan, but remarkable for its brilliant plumage, and active, fly-catching habits. It chooses for its perch the outside twig of a tree, whence it makes its forage amongst the insect tribes that are brought out by the morning beams. This Bee-eater breeds during the months of April and May, laying its eggs in holes in banks. On the 13th May, 1850, I found a pair of these birds breeding in a hole in a bank; the hole was more than an arm's length in depth. At the bottom of it I found three young birds, one very small, with scarcely any feathers on it, another somewhat larger, and the third of considerable size and pretty well fledged. There was no nest."

Mr. Scully shot a specimen in the valley of Nepal in March, and adds (l. c.) that it was common about Hetoura, in the Dun, and in the plains of Nepal near the Tarai in winter. Dr. Fairbank found it abundant at the base of the Palani hills and in the adjacent plains. Vidal records it as abundant at Kelshi and Ratnagiri in South Koukan. Stoliczka obtained it in Southern Kulu in summer; and Mr. Inglis found it very common in North-eastern Cachar, between August and April.

Mr. Cripps says (Str. Feath. vii. p. 258) that it is a "very common and permanent resident (in Eastern Bengal). I have found several of their nest-holes during March and April with from four to five eggs in each. On one occasion I pressed out an egg (without a shell) from the oviduct of a female in the way described by Mr. Adam in 'Nest and Eggs,' p. 101. All the holes I found were on dead level plains, although in one or two instances the high river-banks were close alongside the holes." Mr. Ball (l. c.), in his notes on the birds observed between the Ganges and Godaveri, records it as occurring in the Rajmehal hills, Bardwan, Manbhum, Lohardugga, Singbhum, Sirguja, Sambulpur, north of Mahanadi, Orissa, Nowagarh, and Karial. It ranges down to the extreme south of India and to Ceylon. Mr. Holdsworth (l. c.) found it "exceedingly abundant in the northern part of Ceylon, where it is a resident. It is also found sometimes at Colombo and on other parts of the coast. Whilst living at Aripo, I had constant opportunities of observing these birds closely, as the railings of my veranda were a favourite perching-place for them, and they would allow me to approach within a few feet without showing any alarm. Forty or fifty of these beautiful birds generally roosted in a small bushy tree only a few yards from the house. This species seems to prefer a low station when looking out for its prey, frequently perching on a small stick only a few inches from the ground." Colonel Legge writes ('Birds of Ceylon,' p. 310):—"The Green Bee-eater is a resident species and very numerous in all the dry parts of the low country. It is most abundant about open scrubby land near the sea-coast round the north of the island and along the south-east and eastern sea-boards. Its habitat seems to be restricted to a nicety by the influence of climate. It is common in the interior of the northern half of the island, as well as in the maritime regions, and can be traced along the foot of the western slopes of the Matale ranges from Dambulla to Kurunegala, and thence across the dry country on the north of the Polgahawella and Ambepussa hills to Chilaw and Madampe, near which it stops, not being found south of Nattande. So much does it avoid a moist atmosphere that it extends for a few miles south of Kurunegala, on the highroad to Polgahawella, and suddenly vanishes on the road entering the hills. South of these limits it is unknown throughout the Western Province and the south-west hill-region, reappearing again just to the eastward of Tangalla, where the climate again becomes dry; beyond this all round the coast it is common, being particularly numerous in the Hambantota and Yāla districts. I have traced it through the interior to the foot of the Haputale hills, but it is much scarcer there than at the sea-coast. In the Eastern Province it inhabits the high cheenas in the neighbourhood of Bibile, which attain an altitude of 1000 feet, and which is the highest point I have found it to attain in Ceylon.

Mr. Holdsworth remarks, *loc. cit.*, that it occurs about Colombo. I conclude that the evidence on which this place is included in its range must be that of a stray bird; for I have never observed it anywhere nearer to it than the above limits, neither has Mr. MacVicar nor the taxidermist of the Colombo Museum, both of whom have collected for many years in that part."

Mr. Armstrong (Str. Feath. iv. p. 304) says:—"I found it generally distributed over every part of Southern Pegu which I visited. It was especially abundant at the mouth of the Rangoon river and from there all along the coast up to China Ba-keer, where hundreds might be seen perched upon the dead bushes and drift-wood washed up along the margin of the shore just above high-water mark. They were here wonderfully tame, allowing me to get within two or three yards of them before they would attempt to fly away." According to Mr. Oates (B. of Brit. Burmah, ii. p. 66) this Bee-eater is abundant over the whole of British Burmah except in the south of Tenasserim, where, according to Mr. Davison, it is not found south of Mergui, nor does it in any part of the country ascend the higher mountains.

Mr. Bingham says (Str. Feath. ix. p. 152) that "except in heavy forest-land this little bird is almost as common in Tenasserim as in the N.W. Provinces of India. It crosses the Dawna range into the Thoungyeen valley, and is found in suitable spots all along the river and is a permanent resident, breeding there."

Dr. Anderson (Yunnan Exped. i. p. 582) found it common in the Sunda valley, and remarks that all that were obtained there belonged to the rufous-headed race; but in Cochin China, where this Bee-eater is said to be very common, M. Tiraut remarks that both races are found. It does not appear to occur in China proper.

In habits the Little Green Bee-eater does not differ much from its allies; but it is said to be usually seen in small family parties or in pairs, and never in large flocks, and it is extremely tame and confiding, showing but little fear of man.

Dr. A. Leith Adams, who met with it in Egypt and Nubia, says:—"It is a lively little creature, and on sunny days may be seen sporting about with great vigour, now shooting from the extremity of an acacia branch, anon flitting from furrow to furrow in a newly ploughed field; now four or six are clustered together on a branch, then suddenly, with loud shrieks and chatterings, they break off in divers directions. It is withal a stupid bird, and allows one to approach within a few feet; not even the report of a gun seems to frighten it." Colonel Legge, in his 'Birds of Ceylon,' gives some excellent notes respecting the habits of this Bee-eater, which I transcribe as follows:—"This is one of the most charmingly fearless little birds in Ceylon; unlike the last it is very terrestrial in its habits, perching all day on some little bush or low stick near the ground, and sallying out like a Flycatcher after its food, when it at once returns to its perch or sweeps off to another close by. It is generally found in pairs, or three or four in scattered company, which frequent roadsides and dry open ground of all description where they can find objects to take up their watch upon. About Trincomalie, and, in fact, anywhere on the sea-coast of the eastern side of the island, it is very fond of the sandy scrubby wastes lining the sea-beach, and is so tame that it may be almost knocked down with a stick, so near an approach will it allow before taking wing. In the interior a favourite locality with it is the dried-up paddy-fields in the neighbourhood of the village tanks. It roosts in little colonies, retiring early to rest and congregating in close company; it resorts usually to the same tree, round which much noisy preparation goes on—flying up and wheeling round, alighting on a neighbouring tree-top and then returning, after which the little flock will start out again from the branches and make another little detour, keeping up all the while a continuous clamour. Its note is a sweet little chirrup, unlike the loud voice of the last species. It is either uttered when the bird is perched or when it is sailing along in pursuit of an insect, which it seizes with an audible snap of its bill. It usually preys on small flies or minute Coleoptera, avoiding large dragonflies and other giants of the insect kingdom, upon which the last species feasts and beats to death in the manner aforementioned. Jerdon says that he has seen one occasionally pick an insect off a branch or a stalk of grain or grass; and Blyth has seen them assembled round a small tank seizing objects from the surface of the water, after the manner of a Kingfisher. I have also observed them about rushy jheels and small tanks, but they are not particularly partial to the vicinity of water." Mr. F. Moore also writes (P. Z. S. 1857, p. 87):—"This bird abounds in the neighbourhood of Muttra. Its flight consists of short rapid jerks, and a quick gliding motion, and it generally returns to the same twig from whence it set out. Sometimes several of them may be seen wallowing in the dust on the highroad on a sunny morning. It feeds on insects, and builds its nest in the high banks of the neighbourhood. Its nest is in a very deep horizontal hole in perpendicular banks of hard earth, but often so low as to be within reach of the hand. From this it appears that whilst they guard against other birds, snakes, and squirrels effectually, they fear not man. These nests are generally on the highroad-side, and the birds fly in and out unhesitatingly." I am indebted to Capt. Bingham for the following notes:—" I have dug out dozens of its nests both in the North-west Provinces of India and in Tenasserim, near Moulmein, and also in the interior; for, though not a bird of the heavy forests, wherever there is any open country, especially rice-fields, there it is sure to be found. In the North-west Provinces I have found it breeding from April to July; in Tenasserim in March, April, and May. Sandy banks of rivers and ravines, old mud walls and mounds, sides of roads, deep ruts on them, and even the bare level ground on sandy plains are the sites chosen for its nest-holes, which vary from 2 to 5 feet in depth, and end in rounded chambers about 3 inches in diameter; these I have never found lined in any way, the eggs being laid on the bare ground. Both male and female engage in the work of excavation, digging vigorously with their bills, which get much abraded thereby, and pushing out the loosened sand with a quick backward movement of their feet.

" The following extract from an old note-book of mine well illustrates their perseverance at this task, from which they are not easily driven:—'Allahabad, 10th April. I noticed today a wonderful instance of perseverance in the little Bee-eater (*Merops viridis*) in sticking to its task of digging its nest-hole. A number of these birds have taken possession of the sandy mounds used as butts on the rifle-range belonging to my regiment. This morning when I went down to superintend the individual firing of a party of recruits, previous to the commencement of the practice, I noticed two or three pairs of these little birds busy digging into the face of the butts behind and above the targets. When the firing commenced I distinctly saw two of the Bee-eaters clinging on to the sandy slope hard at work. Now sepoy recruits are not super-excellent shots, and the misses were many, the bullets flying wide of the targets and falling pat-pat-pat on the butt, close and around where the birds were at work. As each shot knocked up its little cloud of dust the birds would flutter off for a foot or so, but immediately in the intervals between the shots obstinately returned and continued digging, tooth and nail. This went on for fully half an hour or longer; at last, after some narrow shaves (it is a wonder none of them were hit) they gave up their task as a bad job and flew off.'

" In Tenasserim, near the large village of Kaukarit, on the Houndraw river, I found last May a nest-hole of this bird dug into the side of a deep cart-rut on a rough road leading across a rice-field. The rut was about a foot or so from a large rock, and the birds had dug straight up to the latter and then tunnelled off at right angles, and slightly downwards, for at least 3 feet, along the face of the rock. The egg-chamber contained one solitary hard-set egg! The usual number of eggs is four, but I have often taken five; they are globular and glossy white in shape and colour. Both sexes, I believe, engage in the work of incubation."

Mr. R. W. Morgan wrote to me that in Southern India he "found this bird breeding on the banks of rivers and ghaut-roads, digging a neat tunnel from three to seven feet in depth, with a globular chamber at the end. The eggs vary from three to six in number, and are deposited on the earth, there being no attempt at a nest." I am indebted to Mr. Morgan for a series of eggs of this Bee-eater, which are pure white in colour, glossy in texture, rather round in form, and in size average about $\frac{4}{10}$ by $\frac{4}{10}$ inch.

The Little Green Bee-eater is subject to considerable variation in coloration of plumage; but after a careful examination of a very large series of specimens I cannot find any valid reason for separating any one of these forms and raising it to specific rank, though at the first glance one would be inclined to consider the extreme form which inhabits Ceylon separable from the Egyptian bird. This latter has the crown sometimes plain green and sometimes tinged with a rufescent golden hue; but the throat is, as a rule, green with a blue streak on each side, though in some specimens I have observed a trace of verditer-blue on the centre of the throat, whereas the form inhabiting Burmah has the head and nape very rufous, but the throat is as in the Egyptian bird. Specimens from Southern India and Ceylon, on the other hand, have the throat verditer-blue, more or less intermixed with green, and the crown and nape tinged with rufous. In the British Museum there is a large series of specimens from various localities, which differ as follows:—Those from Egypt, Nubia, and Abyssinia differ scarcely at all *inter se*, and have the throat green and the head but slightly tinged with rufescent golden. One from Khist, north-east of Bushire, and one from Gwadar, in Baluchistan, have the throat marked with verditer-blue, but not very distinctly, and the head coloured as in Egyptian examples. Two from Madras and one from Lahore have the head rather rufous, that from Lahore being more marked with this colour; but in one there is scarcely any trace of blue on the throat and none in the other two. One from Darjeeling has the crown as in Egyptian specimens, but the throat and underparts are washed with greenish blue; and one from Nepal has the crown tinged with rufous and the throat and underparts more blue than in any other specimen I have examined. Four from Kamptee have the head slightly tinged with rufous, but vary much in the amount of blue on the throat. Lastly, two specimens from Burmah have the head and nape very rufous; but the throat is coloured as in Egyptian specimens, being green with a blue streak on each side.

In the Tweeddale collection there is a very rich series of these Bee-eaters, and on examining them, I find one from Assam and two from Ahmednuggur precisely similar to and undistinguishable from Egyptian examples, whereas three from Rangoon, one from Tonghoo, and one from Central India are like Egyptian specimens, but have the head rather more strongly marked with rufous; a large series from various parts of India (three from Ahmednuggur), Tonghoo, the Karen hills, Moulmein, and Rangoon have blue on the throat, and have the crown and nape rufous, varying much in intensity of colour. In some the blue tinge on the throat is very slight, whereas in others it is nearly as deep as in Ceylonese specimens. In a series from India (Candeish, Maunbhoom, Deyra Doon, Ahmednuggur), Tonghoo, and Ceylon, the throat is verditer-blue, more or less mixed with green, some having the head very rufous, whereas others have it coloured as in Egyptian examples.

Speaking of Ceylonese specimens, Col. Legge writes (B. of Ceylon, i. p. 309) that they "vary in the golden hue of the nape and hind neck, but do not exhibit the brilliant hue of birds from Cachar and Burmah, to which Hodgson gave his name of *ferrugineiceps*: they are typical *M. viridis*, like birds from Central and Southern India; but it must be remarked that occasionally very rufous-headed specimens are procured in Madras. That the species is variable in this character throughout its entire habitat may be gathered from the fact, demonstrated by Mr. Hume, of the Sindh race almost wanting the rusty golden tinge. In Ceylon I have observed that nestling birds vary in the extent of the brighter colours of their plumage when these are first put on, the development of such tints depending perhaps on the physical vigour of the individual. I once shot a pair of young green Bee-eaters together, which were, of course, out of the same nest— one with the normal plain green throat and short tail of the nestling, the other with the blue throat-band appearing and the central tail-feathers half-grown. Perhaps the latter would always have been a more brilliantly plumaged bird than the former; for the difference in age, at most 24 hours, could scarcely have accounted for the backwardness of the plainer specimen in acquiring its adult character. As regards the relative size of Indian and Ceylonese birds, I find that the wings in eight specimens from Pegu (as given in 'Stray Feathers') vary from 3·6 to 3·8 inches, precisely the measurements given above for Ceylonese birds. Some Indian examples have the central tail-feathers longer than any I have seen in Ceylon; one specimen from Kamptee, in the British Museum, has them 2·6 inches beyond the adjacent pair, 2·3 being my limit." To this I may add that, as a rule, I have found that examples from Egypt have the central rectrices longest, and in one specimen in my collection from there, the two central tail-feathers extend slightly over three inches beyond the lateral ones.

The specimens figured are, on one Plate, two adult birds from Egypt, and on the second Plate one from Burmah and one from Ceylon.

In the preparation of the above article I have examined the following specimens :—

E Mus. H. E. Dresser.

a, b, c. Egypt (*Capt. Shelley*). *d.* Abyssinia (*Verreaux*). *e.* India. *f.* Maunbhoom, India, January 1865 (*Beavan*). *g, h.* Ceylon (*Holdsworth*). *i.* Pegu, British Burmah.

E Mus. Tweeddale.

a, ♂ juv. Zoulla, 8th June, 1868 (*Jesse*). *b, c.* India. *d,* juv. India. *e, f, g.* Candeish. *h.* Deyra Doon. *i, ♂.* Huware, near Ahmednuggur, 1st December, 1876. *k, ♀.* Near Ahmednuggur, 2nd January, 1875. *l.* Ahmednuggur, 19th September, 1876. *m.* Ahmeduuggur, 19th October. *n.* Ahmednuggur, December 1876. *o, ♂* ad.; *p, ♂* juv. Ahmednuggur (*Fairbank*). *q, ♂.* Maunbhoom, February 1865 (*Beavan*). *r, ♀.* N. Khasia hills, February 1876. *s, ♀* juv. (*Biddulph*). *t, ♀.* Raiwal Pindee (*Biddulph*). *u, ♂.* Hazaree-bagh (*Biddulph*). *v.* Mysore. *w, ♀.* Khandala, 23rd May, 1876. *x, y.* Ceylon. *z, aa, bb, cc, dd.* Tonghoo. *ee.* Meetan, Tenasserim. *ff.* Assam. *gg.* Burmah. *hh.* Moulmein, 9th July, 1865 (*Beavan*). *ii.* Karen hills (*Wardlaw Ramsay*). *jj.* Rangoon. *kk, ♂.* Rangoon, 21st June, 1873 (*W. Ramsay*). *ll, mm, ♂, ♀.* Rangoon, 29th November, 1873. *nn.* Rangoon, 14th November, 1873 (*W. Ramsay*).

E Mus. Paris.

a. Coromandel coast, type of *Merops citrinella*, Vieill. (*Sonnerat*). *b, c, d, e.* India (*Eydoux & Souleyet*). *f.* Poudicherry (*Leschenault*). *g.* Bengal (*Macé*).

E Mus. Brit.

a, b, c. Egypt (*Sir S. Baker*). *d.* Nubia (*Schaufuss*). *e.* Atfah, Annesley Bay, Abyssinia, 5th February, 1868.

f. Habab, Abyssinia, 9th July, 1868 (*Blanford*). *g, h*. Zoulla, Abyssinia, 8th June, 1868 (*Jesse*). *i*. Mauritius, 24th July, 1838 (*Gould coll.*). *k, ♀*. Khist, N.E. of Bushire (*Major St. John*). *l*. Gwadar, December 1871 (*Blanford*). *m, n*. Madras (*Elliot*). *o*. Lahore (*Marshall*). *p*. Darjeeling (*Hodgson*). *q, r*. Nepal (*Hodgson*). *s, t, u, v*. Kamptee (*Dr. Hinde*). *w*. Moultan (*Capt. Tweedie*). *x*. Cashmere (*Langworthy*). *y*. Kumaon (*Strachey*). *z*. Tenasserim (*Packman*). *aa*. Katch (*E. W. Oates*). *bb*. Lower Pegu, pale canary-coloured variety (*E. W. Oates*). *cc*. Shienpayah, Burmah, 16th January, 1868. *dd*. Mandalay, Upper Burmah, 25th September, 1868 (*Dr. J. Anderson*).

E Mus. E. W. H. Holdsworth.

a, b, ♀. Aripo, N.W. Ceylon, 13th February, 1869. *c, ♀*. Aripo, 12th December, 1869 (*E. W. H. H.*).

E Mus. G. E. Shelley.

a, b, ♂; c, ♀. Egypt, February 1868 (*G. E. S.*). *d, ♂*. Egypt, 15th March, 1868. *e, ♂*. Egypt, 30th March, 1870 (*G. E. S.*).

E Mus. H. J. Elwes.

a, ♂. Bodenar, Kenoor, 7th February, 1870 (*H. J. E.*).

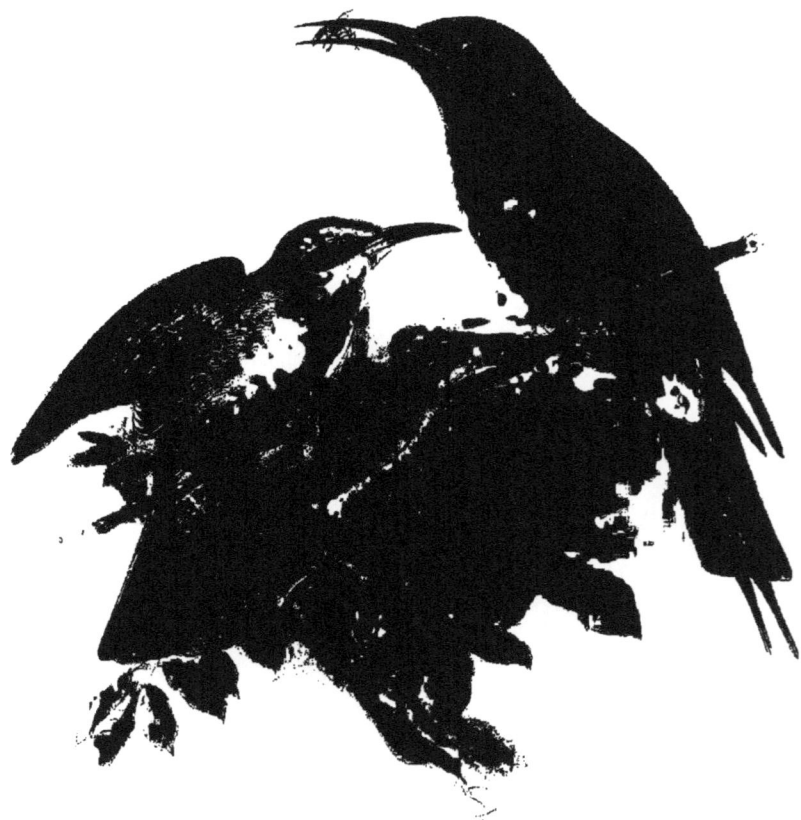

MEROPS CYANOPHRYS.

BLUE-THROATED GREEN BEE-EATER.

Phlothrus cyanophrys, Cab. Mus. Hein. ii. p. 137 (1859–60); Gray, Hand-l. of B. i. p. 100, no. 1212 (1869).
Merops cyanophrys (Cab.), Heugl. Orn. N.O.-Afr. i. p. 203, pl. vi. (1869); Hartlaub, P. Z. S. 1881, p. 957.
Merops cupreus, Ehr. in Mus. Berol.

Figura unica.

Heuglin, Orn. N.O.-Afr. i. pl. vi.

Hab. Arabia, Senaar.

Ad. suprà lætè psittacino-viridis : nuchâ vix rufescenti lavatâ : fronte, superciliis et gulâ pulchrè turcino-cæruleis :
loris et striâ transoculari nigris : corpore subtùs viridi-cærulescente, suprà et subcaudalibus magis cæruleis
remigibus primariis in pogonio externo viridibus, et in pogonio interno rufescenti-cinnamomeis, et nigricanti
fumoso apicatis : secundariis rufescenti-cinnamomeis, extùs virescenti lavatis et nigro apicatis : caudâ viridi,
rectricibus centralibus elongatis.

Juv. sordidior : gulâ viridi-cæruleâ : torque pectorali vix indicato : corpore subtùs sordidè viridi-albido : rectricibus
centralibus nec elongatis.

Adult (Gumfuddc, no. 9389 : type).—Upper parts generally rich parrot-green, slightly washed
with reddish on the nape and the fore part of the back; a narrow frontal line and a tolerably
broad supercilium rich blue; lores and auricular region black; chin and throat brilliant blue, on
the lower part of the throat crossed by a black band; breast and underparts generally blue-green,
fading on the under tail-coverts to dull bluish; primary-quills green on the outer web and dull
reddish on the inner web, tipped with dull blackish; secondaries with the reddish colour brighter
and extending partly over the outer web, all being broadly terminated with black; inner
secondaries and scapulars with the wing-coverts similar to the dorsal feathers; tail square, the
two central rectrices extending nearly an inch beyond the rest; rectrices green, narrowly margined
on the inner web with dull rufous; beak blackish, rather stout at the base and bowed; legs
brownish; iris red. Total length 7½ inches without beak, culmen 1·3, from nostril to tip 1·0,
wing 3·8, tail to tip of central rectrix 4·2, tarsus 0·48.

Young (Gumfuddc, no. 9388).—Resembles the adult bird, but is much paler, the colours,
especially on the upper parts, having a washed-out appearance; the blue supercilium is clearly

G

defined and rich in colour, but the blue and black on the throat are merely indicated, and the rest of the underparts are dirty white washed with blue-green; tail rather uneven, most of the feathers not having reached their proper length; central rectrices only partly grown and shorter than the remainder; bill smaller and weaker than in the specimen above described. Culmen 1·05, from nostril to tip 0·7, wing 3·35, tail 3·1, tarsus 0·48.

UNTIL comparatively lately nothing was known respecting this Bee-eater beyond what was written by Hemprich and Ehrenberg; and the only specimens known were the two above described, obtained by these gentlemen at Gumfudde in Arabia. When in Paris, however, about three years ago, I spent a day in the Museum at the Jardin des Plantes, looking over the specimens which they have in skins packed away in drawers; and just before leaving I found a drawer in which, I was informed, were stowed a lot of bad skins, to be thrown away. These I thought it advisable to look over, and was told that if I found any amongst them that I could make use of, I might have them. After some time I found, to my great delight and utter astonishment, a skin of *Merops cyanophrys*, and of course examined the whole lot most carefully, and was rewarded by finding four more, although badly skinned, yet quite free from moth and entire. Of these I was, thanks to the courtesy of Professor Milne-Edwards, allowed to have two in exchange for some other skins required by the Museum; and on my return to England I had these relaxed and made up into quite good skins. So far as I know, these five specimens (three of which are still in the Paris Museum, and two in my possession) and the two in Berlin are the only examples known to be in Europe at the present time. Hemprich and Ehrenberg obtained the two specimens in the Berlin Museum in the mountains of Gumfudde in Arabia, where von Heuglin says it appears to be common: the five birds which I found in the Paris Museum were obtained by M. Botta at Sennaar, and were labelled by him *Merops viridissimus*; as they were badly made up skins they were evidently thrown aside without being examined and compared with other specimens. Compared with *Merops viridis* the present species is very distinct; the central tail-feathers are much less elongated (the two specimens in my collection having the central rectrices extending 0·6 and 0·9 inch beyond the lateral ones) and but little attenuated; the upper parts are rich deep parrot-green, with the faintest rufous tinge in some specimens only, others having no trace of it; and in both my specimens the frontal blue band is very broad and the underparts are bluish green; in one specimen before me this difference in the coloration of the underparts is very striking, and besides there is no trace of green on the throat.

There is no doubt that this bird is the *Merops cupreus* of Ehrenberg, although Cabanis makes *Merops cupreus* a synonym of *Melittophagus pusillus*; for the types in the Berlin Museum are still marked *M. cupreus*, and, besides, *Merops pusillus* does not occur in Arabia, whence these birds came.

The specimens figured and described are the types in the Berlin Museum, which were, by the courtesy of the late Professor Peters, sent to me so that I could describe them and have them figured.

In the preparation of the above article I have examined the following specimens :—

E Mus. Berol.

a, ad. ; *b*, juv. Gumfudde, Arabia (*Hemprich & Ehrenberg*) : types of *Merops cyanophrys.*

E Mus. H. E. Dresser.

Senaar, July 1839 (*Botta*).

E Mus. Paris.

a, b, c, ad. Senaar, July 1839 (*Botta*).

RUFFHEADED BRAND AND FILER

MEROPS BOHIMI

MEROPS BOEHMI.

BUFF-HEADED GREEN BEE-EATER.

Merops (Melittophagus) boehmi, Reichenow, Orn. Centralblatt, April 1882; id. J. f. Orn. 1882, p. 233.
Merops dresseri, Shelley, Proc. Zool. Soc. 1882, p. 303.

Figuræ notabiles.

Shelley, P. Z. S. 1882, pl. xvi.; Reichenow, J. für Orn. 1882, tab. ii. fig. 3.

Hab. Bumi district, East Central Africa (*Boehm*); Rovuma river, E. Africa (*Thomson*).

Ad. Psittacino-viridis, capite, nuchâ et gulâ rufescenti-cinnamomeis, hâc dilutiore : striâ per oculum ductâ nigrâ et subtùs azureo marginatâ : caudâ et alis viridibus, illâ nigro apicatâ, rectricibus centralibus elongatis et attenuatis, apicibus nigris : remigibus intùs cinnamomeo marginatis et nigro apicatis : subalaribus dilutè cinnamomeis : rostro et pedibus nigris : iride rubrâ.

Adult (Rovuma river).—Crown, upper throat, and nape warm rufous buff, some of the feathers having the tips washed with greenish yellow; upper parts generally parrot-green; wings green, the primary quills terminated with dusky blackish, the secondaries rather broadly tipped with black, except the inner ones, which with the scapulars and upper tail-coverts are slightly washed with blue on the terminal portions; tail green, the two central rectrices much elongated and attenuated, the narrow portion being blackish; rest of the rectrices broadly terminated with black, and narrowly tipped with white on the outer ones, and rather more broadly with blue towards the central pair; a broad black band passes through the eye, and is margined below with turquoise-blue; underparts parrot-green, the abdomen and under tail-coverts washed with blue; under wing-coverts and the basal portions of the inner margins of the quills pale rufescent buff; under portion of the tail greyish brown, the black terminal band showing very distinctly; bill and legs black; iris red. Total length about 9 inches, culmen 1·3, wing 3·1, tail 5·5, central rectrices extending 2·6 beyond the lateral ones, tarsus 0·4.

Obs. The type specimen, which was kindly entrusted to me for examination by the late Dr. Peters of Berlin, agrees with the above, but lacks the elongated central rectrices, and measures—culmen 1·25, wing 3·3, tail 3·0; the third and fourth quills are equal and are the longest, the second is 0·2 shorter than the third, the first 0·95 shorter than the fourth, and the sixth 0·4 shorter than the fifth.

This interesting species, so nearly allied to, yet so thoroughly distinct from, *Merops viridis*, is, up to the present time, known only from two specimens, one of which was obtained by Dr. Boehm in the Bumi district in East Central Africa, and the second sent by Mr. R. J. Thomson to Captain Shelley from the Rovuma river, E. Africa. Curiously enough, the two specimens were received at almost the same time; consequently both were described within a few days of each other, indeed so nearly simultaneously that it became a question whether Dr. Reichenow's name or that given by Captain Shelley should have precedence; but, after making careful inquiries, I have satisfied myself that the former will stand. The specimen obtained by Dr. Boehm lacked the elongated central rectrices, and hence it was by Dr. Reichenow referred, in error, to the genus *Melittophagus*.

The specimen described and figured is the type of *Merops dresseri*, in the collection of Captain G. E. Shelley.

In the preparation of the above article I have examined the only two specimens as yet known, viz. :—

E Mus. Berol.

a. Bumi district, East Central Africa (*Dr. Boehm*): type of *M. boehmi*.

E Mus. G. E. Shelley.

a. Rovuma river, East Africa (*R. J. Thomson*): type of *M. dresseri*.

WHITETHROATED BEE EATER.
MEROPS ALBICOLLIS

MEROPS ALBICOLLIS.

WHITE-NECKED BEE-EATER.

Le Guêpier à gorge blanche ou le Guêpier Cuvier, Levaill. Hist. Nat. Guêp. p. 37, pl. ix. (1807).

Merops albicollis, Vieill. Nouv. Dict. xiv. p. 15 (1816) ; Vieill. Tabl. Encycl. et Méthod. p. 393 (1820) ; Gray, Gen. of B. i. p. 86 (1846) ; Strickland, P. Z. S. 1850, p. 216 ; Bp. Consp. Gen. Av. i. p. 161 (1850) ; von Müll. J. f. Orn. 1855, p. 10 ; Hartlaub, Orn. W.-Afr. p. 39 (1857) ; Schlegel, Mus. Pays-Bas, *Merops*, p. 5 (1863) ; Heuglin, J. f. O. 1864, p. 334 ; Sharpe, Ibis, 1869, p. 192 ; Heugl. Orn. N.O.-Afr. i. p. 201 (1869) ; G. R. Gray, Hand-l. of B. i. p. 99, no. 1209 (1869) ; Blanford, Geol. & Zool. of Abyss. p. 321 (1870) ; Finsch & Hartlaub, Vög. Ost-Afr. p. 186 (1870) ; Finsch, Trans. Zool. Soc. vii. p. 224 (1870) ; Shelley & Buckley, Ibis, 1872, p. 286 ; Sharpe, P. Z. S. 1873, p. 712 ; Ussher, Ibis, 1874, p. 48 ; Reichenow, J. f. O. 1875, p. 18 ; Nicholson, P. Z. S. 1878, p. 355 ; Bocage, Orn. d'Angola, p. 89 (1881).

Merops cuvieri, Licht. Verz. Doubl. p. 15 (1823) ; Lesson, Traité d'Orn. p. 238 (1831) ; Smith, S. Afr. Quart. Journ. 2nd ser. part ii. p. 320 (1834) ; Licht. Nomencl. Av. p. 66 (1854).

Merops savignyi, Swains. Zool. Illustr. ii. pl. lxxvi. (1821-22, nec Aud.).

Aerops albicollis (Vieill.), Reichenb. Meropinæ, p. 82 (1852) ; Cab. Mus. Hein. ii. p. 137 (1859-60).

Figuræ notabiles.

Levaillant, Hist. Nat. Guêp. pl. ix.; Swainson, Zool. Illustr. ii. pl. lxxvi.

Hab. Africa, from Khartoum to Zanzibar, and Senegambia to Angola.

Ad. pileo, fasciâ per oculos ductâ et torque pectorali nigris, hoc subtùs cæruleo marginato : fronte, gulâ et superciliis albis : corpore suprà viridi-cæruleo tincto : nuchâ sordidè aurantiacâ : uropygio et caudâ cæruleis : remigibus rufescentibus extùs viridi tinctis : primariis internis et secundariis nigro terminatis : corpore subtùs albido viridi tincto : rostro nigro : iride coccineâ : pedibus fuscis.

Juv. sordidior : rectricibus centralibus nec elongatis : gulâ sulphureo tinctâ, torque pectorali nigro nec cæruleo marginato.

Adult male (Eylet, Abyssinia).—Crown, a broad stripe through the eye, and a broad band across the throat deep black ; lower nape and fore part of the back apple-green, with a golden orange tinge ; rest of the upper parts green tinged with blue ; quills rufous, externally margined with green, all except the first three or four broadly terminated with black, the elongated inner secondaries being, however, except at the base, almost entirely blue ; tail blue with a faint greenish tinge, the central rectrices much elongated and attenuated ; forehead, a broad band over the eye, and the upper throat white ; the black band across the throat bordered below with blue ; underparts otherwise white, tinged, chiefly on the breast and flanks, with green ; bill black ; feet dull brownish ; iris red. Total length about 11 inches, culmen 1·5, wing 4·2, tail 7·0, central rectrices extending 3·5 inches beyond the lateral ones, tarsus 0·62.

Adult female.—Resembles the male.

Young (Nile below Khartoum).—Differs from the adult in not having the central rectrices elongated, and in being generally duller in colour; the black portions of the plumage have small light edgings to the feathers, the dorsal feathers have paler margins, and the throat is tinged with sulphur-yellow; the black pectoral band is not margined below with blue, only a few of the feathers having narrow blue borders.

This Bee-eater is only found in the Ethiopian Region, ranging from Khartoum and Abyssinia down to Zanzibar on the eastern side of the continent, and Senegambia on the west side down to Angola, but does not occur in South Africa.

According to Heuglin (*l. c.*) it "seems, like most of its allies, to be a migrant or partial migrant in North-east Africa. It is also gregarious, and we observed it in the Bescharin Mountains northwards to 20° N. lat., but on the Nile not above 16°. In the Bogos country, on the Abyssinian and Danakil coasts, in Kordofan and on the White Nile we met with it between July and December. It frequents both the forests and the steppes. In September the males were fresh moulted. According to Brehm this Bee-eater is shy and wary and lives singly; but I met with it exclusively in flocks, which scattered during the day to reassemble in the mornings and evenings. It does not, however, collect in such large flocks as do *Merops superciliosus* and *M. nubicus*. Their call-note is also different from that of their larger allies, being softer and more flute-like."

Messrs. Finsch and Hartlaub say (Vög. Ost-Afr. p. 186) that "with the exception of the southern parts it is found throughout the chief portion of tropical Africa, viz. in Senegal (*coll. Brogden, Latham, Leid. Mus.*), Casamanze (*Verreaux*), Sierra Leone (*Jardine*), Old Calabar (*Jardine*), river Niger (*Forbes*), Gold Coast (*Pel, Weiss*), Grand Bassam (*Grijon*), Gaboon (*Verreaux*), Ogobai river (*DuChaillu*), Angola (*Henderson*), Senaar, south of 16° N. lat. (*Brehm*), Kordofan (*Petherick*), Abyssinian coast-region (*Heuglin*), Winayoore, 180 miles from Massowah (*Daubeny*), Djur and Bongo (*Heuglin*), Blue and White Nile (*Antinori*); in the east observed by Von Heuglin at Tadjura on the Danakil coast of Somali-land, and by Baron von der Decken said to occur in Zanzibar."

Dr. Hartmann (J. f. O. 1866, p. 202) records it as common both on the Blue and White Nile and as found in the gardens of Khartoum, where he obtained a male bird in full nuptial dress in August 1861. Mr. Blanford (Geol. & Zool. of Abyssinia, p. 321) did not find it on the coasts of Abyssinia in December, January, and February, but with some other species it migrated into the country in the spring, and was abundant throughout Samhar in June and July. He observed it as far inland as Rairo in Habab, but it appeared to be restricted to the tropical coast-region. He adds that it usually keeps to trees, but he has seen it settle on the ground in company with *Merops viridis*. According to Mr. Nicholson (P. Z. S. 1878, p. 355) it occurs at Darra-Salam, opposite Zanzibar; and Mr. Sharpe remarks (P. Z. S. 1873, p. 712) that Mr. Wakefield sent several specimens from Mombas, showing that it is not uncommon there. On the west coast Lichtenstein records it from Senegambia. Messrs. Sharpe, Shelley and Buckley, and Ussher record it from the Gold Coast. Messrs. Shelley and Buckley state that it is very common and generally distributed; and Mr. Ussher writes (Ibis, 1874, p. 48) that it is "exceedingly common in every part of Fantee and the Gold Coast. It can always be observed in the vicinity of Cape Coast, especially about bush-paths and hollow roads towards evening, when it may occasionally be seen to collect in large numbers, hawking after insects, and occasionally resting on bushes or low branches of large trees. I have never seen the Bee-eater alone, and should consider it decidedly gregarious. It has no especial peculiarity in its habits to distinguish it."

Dr. Reichenow met with it on the Cameroon river on the river-banks and in the steppes, and DuChaillu and Walker obtained it on the Gaboon. Professor Barboza du Bocage remarks that though included, on the authority of Henderson, in the avifauna of Angola, it has not been obtained there by any of the recent collectors.

In habits this Bee-eater does not appear to differ from its allies. Like them it is gregarious, frequenting river-banks and wooded localities ; and Dr. Reichenow observed it on the Cameroon river wandering about in large flocks, passing high up in the air and remaining in the localities it selects for hunting-purposes in the high trees. According to Dr. Hartmann, it breeds on the Nile in September, making its nest in holes in the river-banks ; and its eggs are, he says, like those of *Merops nubicus*, wax-coloured, unspotted, and measured 21 by 15 mm.

The specimens figured are in my own collection, the adult male being the bird above described.

In the preparation of the above article I have examined the following specimens :—

E Mus. H. E. Dresser.

a, b, c. Eylet, Abyssinia (*Esler*). *d, ♂ .* Kordofan, 11th May. *e, f.* Fantee (*Aubinn*). *g, h.* Fantee. *i.* Connor's Hill, Fantee, 1870 (*Ussher*). *k.* Abyssinia.

E Mus. Brit.

a, ♂ . Rairo, 13th August, 1868 (*Jesse*). *b, ♀ .* Ain, 16th August, 1868 (*Jesse*). *c, ♂ ; d, ♀ .* Rairo, 13th August, 1868 (*Blanford*). *e, ♂ .* Koomaylee, 5th June, 1868 (*Blanford*). *f.* Darfur. *g.* Abeokuta (*Nicholson*). *h,* juv. Nile below Khartoum (*Petherick*). *i, k.* Mombas. *l, m.* West Africa (*Col. Sabine*). *n.* Eylet (*Esler*). *o.* Fantee (*Ussher*). *p.* Gaboon (*Walker*). *q.* Gaboon (*DuChaillu*).

E Mus. Tweeddale.

a, ♂ . Rairo, Abyssinia, 13th August, 1868 (*Jesse*). *b, ♀ .* Koomaylee, 5th June, 1868 (*Jesse*). *c,* juv. Gaboon (*DuChaillu*).

AUSTRALIAN BEE EATER
MEROPS ORNATUS

MEROPS ORNATUS.

AUSTRALIAN BEE-EATER.

Variegated Bee-eater, Lath. Gen. Syn. Suppl. ii. p. 155, pl. 128 (1801).

Merops ornatus, Lath. Ind. Orn. Suppl. p. xxxv (1801); Shaw, Gen. Zool. viii. pt. 1, p. 158 (1811); Steph. in Shaw's Gen. Zool. xiii. pt. 2, p. 73 (1825); Gould, B. of Australia, ii. pl. 16 (1840); Gray, Gen. of B. i. p. 86 (1846); Blyth, Cat. Mus. As. Soc. p. 51 (1849); Reichenb. Vög. Neuholl. i. p. 39 (1849); Bp. Consp. Gen. Av. i. p. 162 (1850); Reichenb. Meropinæ, p. 68 (1852); Macgillivr. Narr. Voy. Rattlcs. ii. p. 356 (1852); Licht. Nomencl. Av. p. 66 (1854); G. R. Gray, P. Z. S. 1859, p. 155; Wallace, Ibis, 1860, p. 147; id. P. Z. S. 1862, p. 338; Schlegel, Mus. Pays-Bas, *Merops*, p. 4 (1863); Rosenberg, J. f. O. 1864, p. 118; Gould, Handb. B. of Australia, i. p. 117 (1865); Finsch, Neu-Guinea, p. 16 (1865); Pelzeln, Novara Reise, Vög. i. p. 50 (1865); Ramsay, Ibis, 1866, p. 366; id. Ibis, 1875, p. 582; Salvadori, Ann. Mus. Civ. Gen. vii. pp. 65, 763 (1875); Salvad. & D'Alb. ibid. p. 814 (1875); Sclater, P. Z. S. 1877, p. 105; Masters, Proc. Linn. Soc. N. S. Wales, i. p. 47 (1877); Castlenau & Ramsay, tom. cit. p. 379 (1877); Ramsay, tom. cit. p. 589 (1877); D'Alb. Sydn. Mail, 1877, p. 248; id. Ann. Mus. Civ. Gen. x. p. 12 (1877); Salvadori, Atti R. Ac. Sc. Tor. xiii. p. 312 (1878); Ramsay, Proc. Linn. Soc. N. S. Wales, ii. p. 179 (1878); Forbes, P. Z. S. 1878, p. 121; Tweeddale, Trans. Zool. Soc. viii. p. 42 (1878); Ramsay, Proc. Linn. Soc. N. S. Wales, iii. p. 263 (1879); Sharpe, Journ. Linn. Soc., Zool. xiv. p. 686 (1879); Meyer, Ibis, 1879, p. 57; Layard, Ibis, 1880, p. 300; Ramsay, Proc. Linn. Soc. N. S. Wales, vii. p. 21 (1883).

Guêpier Thouin ou à longs brins, Levaill. Hist. Nat. Guêp. p. 26, pl. 4 (1807).

Philemon ornatus (Lath.), Vieill. Nouv. Dict. xxvii. p. 423 (1818); id. Tabl. Encycl. et Méthod. p. 614 (1822).

Mountain Bee-eater, Lewin, B. of New Holland, pl. 18 (1821).

Merops melanurus, Vig. & Horsf. Trans. Linn. Soc. xv. p. 208 (1826); Lesson, Man. d'Orn. i. p. 87 (1828); id. Traité d'Orn. p. 238 (1831).

Merops thouini seu tenuipennis, Dumont, Dict. des Sc. Nat. xx. p. 52, éd. Levrault (1821).

Merops thouinii, Müll. Verh. Land- en Volkenk. p. 138 (1839–44).

Merops cærulescens, Gray, Gen. of B. i. p. 86 (1846).

Melittophagus ornatus (Lath.), Reichenb. Meropinæ, p. 82 (1852).

Cosmaerops ornatus (Lath.), Cab. Mus. Hein. ii. p. 138 (1859–60); Gray, Hand-l. of B. i. p. 100, no. 1217 (1869).

Cosmaerops cærulescens, Gray, Hand-l. of B. i. p. 100, no. 1218 (1869).

Merops modestus, Oustalet, Assoc. Sc. de France, Bull. n. 533, p. 248 (1878); Salvadori, Atti R. Ac. Sc. Tor. xiii. p. 312 (1878).

Figuræ notabiles.

Gould, B. of Australia, ii. pl. 16; Levaillant, Hist. Nat. Guêp. pl. 4; Reichenbach, Meropinæ, pl. ccccxlvi. figs. 3233, 3234.

HAB. Papuasia, Moluccas, New Guinea, Australia.

♂ *ad.* pileo et nuchâ aurantiaco-fuscis, illo vix viridi tincto : fronte et lineâ superciliari, dorso et tectricibus alarum viridibus fusco tinctis : alis aurantiaco-fuscis conspicuè nigro apicatis, pennis extùs viridi lavatis ; secundariis intimis elongatis, viridibus cæruleo tinctis : dorso imo, uropygio et supracaudalibus cærulcis : caudâ nigrâ vix furcatâ, rectricibus duabus centralibus valdè elongatis attenuatis et indistinctè spatulatis : vittâ per oculum ductâ et regione paroticâ intensè nigris, subtùs turcino marginatis : gulâ flavâ, lateraliter aurantiaco lavatâ : torque pectorali nigro : corpore subtùs viridi fusco tincto : subcaudalibus cærulcis : iride fusco-rubrâ : rostro nigro : pedibus viridi-cinereis.

Juv. suprà sordidè viridis cæruleo tinctus : uropygio, supracaudalibus et secundariis intimis sordidè cærulcis : corpore subtùs sordidè viridi-cæruleo : gulâ sordidâ ochraceâ : torque pectorali nullo : alis et caudâ sicut in adulto, sed sordidioribus : rectricibus centralibus nec elongatis.

Adult male (Port Moresby).—Crown and nape orange-brown, tinged with green on the fore part of the crown ; forehead and a line over the eye, back, and wing-coverts green tinged with brown ; wings orange-brown, broadly tipped with black; the quills externally washed with green, the elongated inner secondaries green tinged with blue ; lower back, rump, and upper tail-coverts cærulean blue ; tail black, slightly forked, the two central rectrices considerably elongated, attenuated, and slightly spatulated ; lores and a broad band through the eye and the ear-coverts rich velvety black, bordered below with turquoise-blue ; throat bright yellow, becoming orange on the sides of the neck and lower throat; below this the breast is crossed by a broad black band ; underparts green tinged with brown ; under tail-coverts bright cærulean blue ; iris light brownish red ; bill black ; legs greenish grey. Total length about 9 inches, culmen 1·7, wing 4·4, tail 5·2, central rectrices extending 1·78 beyond the lateral ones, tarsus 0·45.

Immature (Cape York).—Much duller in general coloration than the adult, the nape much greener, the blue line beneath the eye paler, the pectoral band very slightly marked, and the two central rectrices only extending a little beyond the lateral ones.

Young (Moluccas).—Upper parts dull greenish tinged with blue ; the rump, upper tail-coverts, and inner secondaries dull cærulean ; underparts dull greenish blue, the throat pale dull ochreous, no blue line below the eye or black pectoral band ; wings and tail as in the adult, but duller, and the central rectrices not elongated.

Obs. Specimens which I have examined vary somewhat in intensity of coloration and especially as regards the amount of blue in the plumage. A specimen in the Tweeddale collection from Port Albany has the crown, scapulars, and back slightly tinged with turquoise-blue ; and one in the British Museum, a female from Dorey, has the abdomen washed with blue and is labelled *Merops cærulescens*. Another example in the British Museum, a female from Lomboek, is peculiar in having a tolerably broad band of blue below the black pectoral band, but does not otherwise differ from typical examples of this species. From an examination of the specimens I can indorse Mr. Wallace's remarks (*l. c.*) that specimens from the Sula Islands agree with those from Ternate in having more brown on the head and less blue on the breast than the Timor and Lomboek specimens ; and according to Mr. Ramsay (Ibis, 1866, p. 326) " specimens from Port Denison are somewhat smaller than those from New South Wales."

THE present species ranges from Borneo, Java, and the Moluccas to New Guinea and the whole of Australia. There is a specimen from Java in the Tweeddale collection. Wallace obtained it at Celebes, Flores, Lombock, Timor, Sula Island, Sumbawa, Ternate, and Mysol, and both he and other collectors obtained it in New Guinea. Bernstein records it from Gilolo, and Gould and other authors record it from various parts of Australia and New South Wales.

Ramsay (Proc. Linn. Soc. N. S. W. ii. p. 179) gives its range in Australia as Port Darwin, Port Essington, Gulf of Carpentaria, Cape York, Rockingham Bay, Port Denison, Wide Bay district, Richmond and Clarence River district, New South Wales, the interior of Australia, Victoria, South Australia, West Australia, and the south coast of New Guinea; and Salvadori (Orn. della Papuasia, i. p. 401) records it from Torres Straits, New Britain, Duke of York Island, Aru, New Guinea, Fly River, Dorey, Andai, Mansinam, Tarawai, Sorong, Salwatti, Mysol, Jobi, Mysori, the Moluccas, Halmaheira, Ternate, Batchian, Buru, Ceram, Amboina, Sula, Celebes, Timor, Sumbawa, Flores, Lombock, and Java.

Dr. Meyer says (Ibis, 1879, p. 57) that "in the Minahassa it is only numerous in the east monsoon. Near Menado in May, and on the Togian Islands in August. As to the development of the lengthened tail-feathers, an examination of a series of specimens proves that, as in the analogous case of *Prioniturus*, the lengthened tail-feathers are narrower *ab initio*, and are not formed by being rubbed off, except at the last stage, which, however, does not touch the principle that also here immanent causes affect the shape of these feathers. The same remarks apply to *M. philippinus.*"

Rosenberg (J. f. O. 1864, p. 118) records it from "Amboina, Ceram, Timor, and North Australia;" and Gould (Handb. B. Austral. i. p. 117) writes respecting its occurrence in Australia that "it arrives in New South Wales and in all the colonies lying within the same degree of latitude in August, and departs in March, the intervening period being employed in the duties of incubation and of rearing its progeny. During the summer months it is universally spread over the whole southern portion of the continent from east to west, and in winter the northern. In South Australia and at Swan River it is equally numerous as in New South Wales, generally giving preference to the inland districts rather than to those near the coast; hence it is rarely to be met with in the neighbourhood of Perth, while in the York district it is very common. In New South Wales I found it especially abundant on the Upper Hunter, and all other parts towards the interior, as far as I had an opportunity of exploring." Mr. E. L. Layard obtained it in New Britain, but did not observe any on the Duke of York Islands.

The best account I find respecting the habits of this bird is that published by the late Mr. Gould (Handb. B. of Austral. i. p. 118) as follows:—"Its favourite resorts during the day are the open, arid, and thinly-timbered forests; and in the evening the banks and sides of rivers, where numbers may frequently be seen in company. It almost invariably selects a dead or leafless branch whereon to perch, and from which it darts forth to capture the passing insects. Its flight somewhat resembles that of the *Artami*, and although it is capable of being sustained for some time, the bird more frequently performs short excursions, and returns to the branch it had left.

"The eggs are deposited and the young reared in holes made in the sandy banks of rivers or any similar situation in the forest favourable for the purpose. The entrance is scarcely larger than a mouse-hole, and is continued for a yard in depth, at the end of which is an excavation of sufficient size for the reception of the four or five pinky-white eggs, which are ten lines long by eight or nine lines broad. The stomach is tolerably muscular, and the food consists of various insects, principally Coleoptera and Neuroptera."

The specimens figured and described are in my own collection.

In the preparation of the above article I have examined the following specimens :—

E Mus. H. E. Dresser.

a, b, ad. ; *c,* juv. Cape York (*Cockerell*).   *d,* ad. Queensland (*Whitely*).   *e,* ad. Port Moresby (*Gerrard*).   *f,* juv. Moluccas.

E Mus. Brit.

a. Australia (*Jukes*).   *b.* S. Australia (*Sturt*).   *c, d, ♂ ♀* . Moreton Bay (*Gould*).   *e, ♂ ; f, ♀* . Cape York (*' Challenger' Expedition*).   *g, h.* Cape York (*Cockerell*).   *i.* Cape Upstart, Australia.   *k.* S. Australia.   *l.* Queensland (*Nicholson*).   *m.* New South Wales (*Gould*).   *n, o, ♀* . Booby Island (*' Challenger' Expedition*).   *p.* Point Pearse. *q.* Sula Island (*Wallace*).   *r, s, ♂* . Lomboek (*Wallace*).   *t.* Lombock, 1856 (*Wallace*).   *u.* Menado, North Celebes, 1859 (*Wallace*).   *v,* ad.; *w,* juv. Mysol (*Wallace*).   *x.* N. Ceram (*Wallace*).   *y, ♂ ; z, ♀* . Dorey, 1858 (*Wallace*).   *aa.* Menado, N. Celebes, 1859 (*Wallace*).

E Mus. Tweeddale.

a. Port Albany, N. Australia.   *b.* South Queensland.   *c,* juv. Ceram (*Wallace*).   *d, e.* Menado, N. Celebes. *f.* Java.   *g.* Sula Island (*Wallace*).

BLUE TAILED BEE EATER
MEROPS PHILIPPINUS

MEROPS PHILIPPINUS.

BLUE-TAILED BEE-EATER.

Apiaster philippensis major, Briss. Orn. iv. p. 860, pl. xliii. fig. 1 (1760).

Grand Guépier des Philippines, D'Aubenton, Pl. Enl. no. 57.

Le Guépier vert à queue d'azur, Montb. iu Buff. Hist. Nat. Ois. vi. p. 504 (1779).

Merops philippinus, Linn. Syst. Nat. ed. 13, i. p. 183, no. 5 (1767); Gmel. Syst. Nat. i. p. 461 (1788); Shaw, Gen. Zool. viii. pt. 1, p. 165 (1811); Vieill. Nouv. Dict. xiv. p. 27 (1817); Franklin, P. Z. S. 1831, p. 115; Ewer, P. Z. S. 1842, p. 92; Strickland, P. Z. S. 1842, p. 167; Gray, Gen. of B. i. p. 86 (1846); Horsf. & Moore, Cat. B. E.I. Co. Mus. i. p. 86 (1854); Licht. Nomencl. Av. p. 66 (1854); F. Moore, P.Z. S. 1854, p. 263; Gould, B. of Asia, part vii. (1855); F. Moore, P. Z. S. 1857, p. 87; Cassin, U.S. Expl. Exp. p. 228 (1858); Gould, P. Z. S. 1859, pp. 150, 151; Cab. Mus. Hein. ii. p. 139 (1859); Irby, Ibis, 1861, p. 228; Sclater, P. Z. S. 1863, p. 213; Schlegel, Mus. Pays-Bas, *Merops,* p. 2 (1863); Hume, Nests & Eggs of Ind. B. p. 101 (1873); Holdsworth, Ibis, 1874, p. 125; Salvadori, Ucc. Born. p. 89 (1874); Legge, Ibis, 1875, p. 281; Fairbank, Str. Feath. iv. p. 254 (1876); Hume, Str. Feath. iv. p. 287 (1876); David & Oustalet, Ois. de la Chine, p. 78 (1877); Tweeddale, P. Z. S. 1877, pp. 540, 690; Ball, Str. Feath. v. p. 413 (1877); Fairbank, Str. Feath. v. p. 394 (1877); Tweeddale, P. Z. S. 1878, pp. 107, 282, 340, 709; Hume, Stray Feath. vi. p. 498 (1878); Ball, Str. Feath. vii. p. 203 (1878); Cripps, Str. Feath. vii. p. 258 (1878); Tiraut, Bull. Com. Agr. de la Cochin Chine, sér. 3, vol. i. p. 98 (1879); Hume, Str. Feath. viii. p. 48 (1879); Bingham, Str. Feath. viii. p. 198 (1879); Butler, Str. Feath. viii. p. 386 (1879); Meyer, Ibis, 1879, p. 57; Sharpe, Ibis, 1879, p. 248; Bingham, Str. Feath. ix. p. 152 (1880); Vidal, Str. Feath. ix. p. 49 (1880); Legge, B. of Ceylon, p. 306 (1880); Nicholson, Ibis, 1881, p. 143; Kelham, Ibis, 1881, p. 378; Oates, B. of Brit. Burm. ii. p. 66 (1883).

Le Guépier à queue d'azur ou le Guépier Daudin, Levaill. Hist. Nat. Guêp. p. 49, pl. 14 (1807).

Merops philippensis, Steph. in Shaw's Gen. Zool. xiii. pt. 2, p. 75 (1825); Jerdon, B. of Ind. i. p. 207 (1862); Swinhoe, Ibis, 1865, p. 230; Beavan, Ibis, 1865, p. 407; id. Ibis, 1867, p. 318; Swinhoe, Ibis, 1870, p. 91; Legge, Ibis, 1874, pp. 13, 125; Ball, Str. Feath. ii. p. 386 (1874); id. op. cit. iii. p. 289 (1875); Hume, Str. Feath. iii. p. 456 (1875); id. op. cit. v. p. 18 (1877); Anderson, Yunnan Exped. i. p. 511 (1878).

Merops daudini, Cuvier, Règne Anim. i. p. 442 (1829, ex Levaill.).

Merops javanicus, Horsf. Trans. Linn. Soc. xiii. p. 171; Steph. in Shaw's Gen. Zool. xiii. pt. 2, p. 73 (1825); Lesson, Man. d'Orn. ii. p. 86 (1828); Eyton, P. Z. S. 1839, p. 101; Gray, Gen. of B. i. p. 86 (1846); Bp. Consp. Gen. Av. i. p. 162 (1850); Reichenb. Meropinæ, p. 66 (1852); Motley, P. Z. S. 1863, p. 213; Pelzeln, Novara Reise, Vög. i. p. 50 (1865).

Merops javanica (Horsf.), Swains. Classif. of B. ii. p. 333 (1837).

Merops typicus, Hodgson in Gray's Zool. Misc. p. 82 (1844).

Merops cyanorrhos, Temm. in Mus. Lugd. fide Bp. Consp. Gen. Av. i. p. 162 (1850).

Merops savignyi, in Mus. Massen. fide Cab. Mus. Hein. ii. p. 139.

Blepharomerops javanicus (Horsf.), Reichenbach, Meropinæ, p. 82 (1852).

Blepharomerops philippinus (Gmel.), Reichenb. ut suprà (1852); G. R. Gray, Hand-l. of B. i. p. 99, no. 1207 (1869).

Blepharomerops daudini (Cuv.), G. R. Gray, Hand-l. of B. i. p. 99, no. 1208 (1869).

Merops daudini, Cuv., Swinhoe, P. Z. S. 1871, p. 349; Hume, Str. Feath. ii. pp. 102, 469 (1874); Oates, Str. Feath. iii. p. 49 (1875); Armstrong, Str. Feath. iv. p. 304 (1876).

Merops philippensis (Steph.), Hancock, Cat. of B. of North. & Durh. p. 28 (1874).

Figuræ notabiles.

Gould, B. of Asia, part vii.; D'Aubenton, Pl. Enl. no. 57; Levaillant, Hist. Nat. Guêp. pl. 14; Reichenbach, Meropinæ, pl. cccexliv. fig. 3227.

Boro-putringa, Beng.; *Burra-putringa*, Hind.; *Komu passeriki*, Tel. (Jerdon); *Kachangan*, Java; *Burong langir, Berray Berray*, Malay; *Shale*, Nicobarese; *Kurumenne kurulla*, lit. "Beetle-bird," Sinhalese; *Kattalan kuruvi*, lit. "Aloe-bird," Tam.; *Pappugai de Champ*, Portug., lit. "Ground-Parrot" (*apud* Layard).

HAB. Indo-Malayan region, ranging eastward to China and the Philippines and southward to Java.

Ad. suprà saturatè viridis vix fusco lavatus, pileo et nuchâ fusco-olivaceis: uropygio et supracaudalibus saturatè viridi-cœruleis: secundariis intimis eodem colore tinctis: remigibus nigricanti tinctis; scapis nigris: caudâ saturatè cœruleâ vix viridi tinctâ: rectricibus duabus centralibus elongatis versus apicem nigricantibus: mento flavo, gulâ castaneo-rufâ: loris et vittâ magnâ per oculum ductâ nigris, hâc subtùs turcino-cœruleo marginatâ: corpore subtùs pallidè viridi, lateraliter cinnamomeo lavato: subcaudalibus pallidè cœruleis: alis subtùs cinnamomeis: rostro nigro: pedibus nigricantibus: iride rubrâ.

Juv. suprà viridior: plumis ad basin fusco-viridibus: uropygio, supracaudalibus et caudâ sordidioribus: rectricibus centralibus haud elongatis, vittâ nigrâ in facie lateraliter indistinctè cœruleo marginatâ: mento pallidè et sordidè flavo: gulâ pallidè cinnamomeâ: iride sordidè rubrâ.

Adult male (Ceylon).—Upper surface of the head, body, and wings deep green, with a slight brownish tinge, the crown and nape becoming brownish olivaceous; rump and upper tail-coverts deep greenish blue, and the inner secondaries tinged with the same colour; quills tinged with black and having black shafts; tail deep blue with a greenish tinge, the terminal portion of the central rectrices blackish; chin yellow; throat rufous; a broad black streak, margined below with pale turquoise-blue, passes along the side of the head through the eye; underparts of the body pale greenish, becoming pale cœrulean blue on the under tail-coverts; flanks tinged with cinnamon-rufous, which colour pervades the under surface of the wings; bill black; legs and feet blackish; iris scarlet, eyelids grey; mouth flesh-colour. Total length about 11 inches, culmen 1·8, gape 2·0, wing 5·2, tail 6·0, central rectrices 2·2 longer than the lateral ones, tarsus 0·5, middle toe with claw 0·8.

Adult female.—Does not differ from the male in plumage, but appears, as a rule, to have the central rectrices rather shorter than in the male.

Young (Malacca).—Upper parts much greener than in the adult, the feathers brownish green on the basal portions; rump, tail-coverts, and tail much duller than in the adult, the central rectrices not elongated; the blue streaks on the sides of the head scarcely perceptible; the chin much less yellow than in the adult, and the throat pale dull cinnamon instead of rufous; iris dull reddish.

THIS, one of the commonest of the Asiatic Bee-eaters, is found throughout the whole peninsula of India, ranging south to Ceylon and the Andaman Islands; it also inhabits Burmah, Siam, Cochin China, China, the Malay peninsula, Java, Sumatra, Borneo, Celebes, and the Philippine Islands.

According to Dr. Jerdon (*l. c.*), it is "spread more or less over all India and Burmah, extending to Ceylon in the south, and to the Malay peninsula and islands in the east. It prefers forest-countries and well-wooded districts, and, though generally spread, is yet somewhat locally distributed, and you may pass over considerable tracts of country without meeting one. The Malabar coast is always a favourite haunt, and this Bee-eater appears to prefer the neighbourhood of water. It is sometimes found in the Wynaad and other elevated regions of Malabar, but in general prefers a low level." In North-west India it is found as far as Sindh, and Mr. Scrope-Doig met with it in the Narra district, where he believes it breeds; but west of Sindh it seems to be entirely replaced by *Merops persicus*. Mr. A. O. Hume received it from the Mount Aboo district, where Dr. King also obtained it. It is found very generally throughout Central and Eastern India during the cool season. Col. Irby met with it in Oudh and Kumaon in the hot season, but not in any numbers. Mr. Ball records it from Lohadugga, Sirguja, Sambalpur, &c., from the Ganges to the Godavery, in Jaipur and Raipur, but he adds that it is very rare in Chota Nagpur, where he never observed it, however, before the hot weather. Dr. Fairbank, who procured it in the Palani Hills, writes (Str. Feath. v. p. 394) that it "was common in 1866 on the eastern side of the Palani Hills at 2000 or 3000 feet; but this year I only saw it once, and the one I shot, falling among high grass, was not recovered, though I carefully marked the spot where it fell." There are numerous records of its occurrence in various parts of India down to the southern part of the peninsula, too numerous to cite in detail. I may, however, remark that, according to Blyth, it occurs in Lower Bengal chiefly or only during the rainy season. Mr. Cripps says (Str. Feath. vii. p. 168) that it is "far from rare at Furreedpore, Eastern Bengal, where they appear in February, breed in holes in banks in July and August, after which they disappear. They frequent river-banks and 'beels:' in the latter they perch on the sticks and bamboos which the fishermen put down for drying their nets on. They have a much louder note than *Merops viridis*, and are rather shy." Col. Legge writes (*l. c.*) that it is "migratory to Ceylon, arrives in the north of the island about the beginning of September, and rapidly spreads more or less through all parts of it before the end of the month. It seems to find its way to the south-west corner, or Galle district, almost as soon as to any part of the island, and collects there in greater numbers than elsewhere on the western side. I have met with it in the interior of the country, between Galle and Akurresse, as early as the 8th of September. It locates itself in great numbers in the Jaffna peninsula, and on the north-west coast as far south as Puttalam, and spreads in tolerable numbers into the interior, passing over the forest-clad portions, however, to a great extent, and ascending to the patnas and open hills of the Kandyan Province. In Uva and Pusselawa, and on the Agra, Lindula, and Bopatalawa patnas, at an elevation of 5000 feet, it is common; but I have never seen it on the 'plains' of the Nuwara-Elliya plateau. In the Eastern Province it confines itself mostly to the sea-board, being less numerous in the Park country and the south-eastern 'jungle-plain' than the next resident species (*Merops viridis*). Its departure from the island is as sudden as it is regular, in proof of which I may state that at Galle, in two successive seasons, I observed it collect in large flocks between the 29th and 31st March, and disappear entirely on the 1st April. Mr. Holdsworth, who writes that at Aripu it was so abundant that the common resident species (*Merops viridis*) was scarce in comparison with it, states that it left about the beginning of April; and by the end of that month, I believe,

I

it has quitted the island entirely. In the neighbourhood of Colombo it is chiefly located in large tracts of paddy-ground and about the great swamp there and Negombo. It is now and then met with in the cinnamon-gardens."

Davison says (Str. Feathers, ii. p. 162) that "it occurs in the Nicobars, but not in the Andaman group, where it is replaced by *M. quinticolor.* In habits it differs much from both *M. viridis* and *M. quinticolor;* it may be seen for an hour at a time, taking long sailing flights. I have seen ten or twelve of these birds hawking over the grassy hills in the interior of the island of Camorta. I was unable to ascertain whether they breed in the Nicobars or not. It is known to the Nicobarese by the name of Shale. I saw it at the Cocos, but failed to procure a specimen."

Subsequently, however, to when the above notes were written, it has been recorded by Mr. Hume (Str. Feath. iv. p. 287) as having been obtained by Roepstorff at Aberdeen, S. Andamans, in November. It appears to be somewhat rare in Tenasserim, but common in Pegu and Burmah. Mr. Oates says (Str. Feath. iii. p. 49) that it "occurs in large flocks all over Upper Pegu, and is a constant resident. It is, however, very uncertain in its movements, and appears to be locally migratory. In the rains there are comparatively few, and these are seen singly in the paddy-fields perching on bushes. It breeds in all the large nullahs with steep banks, and I lately came across a large colony in the Irrawaddy; but I have hitherto failed to meet with the large colonies mentioned by Jerdon. It occurs nearly to the summit of the Pegu Hills; but I did not find it on the eastern slopes. It occurs again in the plains near Tonghoo." To this Capt. Feilden adds that "it breeds in vast numbers on the banks of the Irrawaddy; the young leave the nest at the beginning of the rains."

Mr. Armstrong, who met with this species in the delta of the Irrawaddy, writes (Str. Feathers, iv. p. 304) that "though tolerably abundant in certain localities, it was by no means general in its distribution. I have only met with it in a tidal swamp a few miles from Elephant Point, and also along the course of Deserters' Creek. In this latter locality it was met with in tolerable abundance, more especially where the margins were bordered with tall *Sonneratia* trees. Here numbers of the species might be seen making wide circles with a strong rapid flight at a great height up in the air and again returning to perch on the summits of these trees, where they would remain for a moment or two before starting on a fresh expedition. They kept, as a rule, to the highest trees, and were very wary and difficult to approach."

Capt. Bingham writes (*l. c.*) as follows:—"In March 1877, I found large parties of this Bee-eater breeding in the sandy banks of the Salween at Shwaygoon. It is not uncommon, and breeds at Kaukarit, on the Houndraw. I observed a pair or two there as late as June the 29th." Again (Str. Feath. ix. p. 152) he writes: "This bird, being partially migratory, is often overlooked; but it is common nearly all the year round at Kaukarit, on the Houndraw river, where it breeds in April and May in the sandy banks of the Kaukarit Choung. In the Thoungyeen valley I have procured it at Meeawuddy in June, at Laidawgyer in April, and on the Dawna Pass in November." According to Mr. Oates (B. of Brit. Burmah, ii. p. 67) it is "found over the whole of British Burmah, being very abundant in Arrakan and Pegu and somewhat rare in Tenasserim. It appears to be partially migratory in Tenasserim; but in Pegu I have observed it during the greater portion of the year, and I think it is resident." It is found in China and Cochin China, and Tirant states (*l. c.*) that it is "very common in Cochin China in all wooded localities." Strickland records it from Canton. Messrs. David and Oustalet state that it visits Southern China in the summer season; and Swinhoe procured it at Swatow, and says that it is supposed to breed in the Wenchang district, N.E. Hainan.

It appears to be tolerably common in Siam and on the Malay peninsula. Dr. Meyer writes ('Ibis,' 1879, p. 57) that "In the Minahassa this bird is only plentiful at certain times, viz. in the dry season during the east monsoon; in the west monsoon it is rarely to be met with. In Limbotto I got it in July, in Makassar in October, 1871, later in Singapore, in December 1871, on Luzon in February 1872, on Negros (Philippine Islands) in March 1872." I have examined specimens from Java, Sumatra, Borneo, and the Celebes, which agree closely with others from India, Burmah, and the Philippine Islands; and Lord Tweeddale remarked that "specimens from the Philippines in no respect differ from Luzon and Negros individuals, or, indeed, from examples from any part of the Indian region," and he adds that it has not hitherto been recorded from Zebu. Mr. Mottley writes (P. Z. S. 1863, p. 213) that in Borneo "it is a very common bird in open places, sailing in circles to hunt the larger Coleoptera and Hymenoptera. It also makes great havoc among the dragonflies with which the air is sometimes filled here. When these birds have seized their prey, they return to their stand, usually a bare high branch, and there kill it by beating it against the twigs. Great numbers of them may sometimes be seen together in the evening flying in one direction, uttering the cry (*pink-pink*) which gives their name."

Before concluding the notes on the range of this Bee-eater, I must remark that it has been included in the British list; but I think there must be some mistake in the matter. Mr. Hancock writes (B. of Northumb. & Durh. p. 28) that he examined a specimen, belonging to the Rev. T. M. Hicks, which was shot near the Snook, Seaton Carew, in August 1862, by Mr. Thos. Hann, of Byer's Green. Thanks to the courtesy of Mr. Hicks, I have had an opportunity of examining this bird and comparing it with my specimens, and I also took advantage of this opportunity to exhibit it at a meeting of the Zoological Society. There is no doubt that the specimen is an old bird of *Merops philippinus* in very perfect plumage; but how a bird with the power of flight of the present species, which has not been hitherto observed west of Sindh, can have come alive over to our island, I cannot imagine: it may have been "changed at nurse" when in the possession of the bird-stuffer, for it is well known that, owing to a similar mistake, the American Waxwing has also been erroneously included in the British list.

Speaking of the habits of this Bee-eater, Dr. Jerdon says (*l. c.*) that "it is mostly observed in scattered parties, perching on high trees, often among paddy-fields, and it in general takes a much longer circuit than *Merops viridis* before returning to its perch. I have often seen one seated on a low palisade, or stump of a tree overhanging a nullah or back-water, every now and then picking an insect off the surface of the water.

"They feed on wasps, bees, dragon-flies, bugs, and even on butterflies, which I have seen this species frequently capture. The flight of this Bee-eater is very fine and powerful, now dashing onwards with rapid strokes, and a velocity that can beat that of a dragon-fly, having captured which, it flaps along with more measured time, now and then soaring with outspread wing. The voice is a full mellow rolling whistle. On one occasion I saw an immense flock of them, probably many thousands, at Caroor, on the road from Trinchinopoly to the Nilghiris; they were perched on the trees lining the fine avenue there, and every now and then sallied forth for half an hour or so, capturing many insects, and then returning to the trees. These birds were probably collected there previous to migrating to their breeding-quarters. They nestle, like *Merops viridis*, in holes in banks of large rivers. I have not seen in India any of these breeding-haunts, but I have lately seen them breeding in thousands on the banks of the Irawaddy in Burmah, in April and May. It would be interesting to know if all the birds of this species that spread themselves over Southern India in the cold weather retire to the wooded banks of this noble river to breed. Mr. Philipps, however, mentions that he found this species breeding ' in on old rampart opposite my house' at Muttra, in the North-western Provinces; and it probably nestles in the banks of the Ganges and Jumna, though I have seen no record to that effect." Mr. Holdsworth, who met with it in Ceylon, says that it is "a noisy bird, with a lofty, dashing flight, successfully pursuing the dragonflies, and then sailing back on outstretched wings to its favourite station on the dead branch of some neighbouring tree, where the insect is killed and swallowed. In the early mornings of March, when there has been but little wind stirring, and the sea was as smooth as glass, I have frequently observed these Bee-eaters hunting for insects close to the surface, and a quarter of a mile from the shore." Col. Legge (*l. c.*) writes that it "prefers to frequent open lands, plains studded with bushes near the sea-shore, esplanades, paddy-fields, swamps, and the patnas of the hill-region. It passes a great part of its existence on the wing, in pursuit of insects, after which it dashes with a very rapid flight, constantly uttering meanwhile its loud notes. When reposing from its labours, it rests on low objects, such as stumps of trees, fences, low projecting branches, little eminences on the ground, and often on the level earth itself. It is tame in its nature, allowing a near approach before it takes wing. On rainy evenings in November and December, when the air is swarming with insects, and particularly with winged termites, which issue forth from their nests on such occasions, the Blue-tailed Bee-eater congregates in large flocks on the wing, dashes to and fro for hours together, ascending to a great height in pursuit of its prey, and keeping up its not unpleasant notes without intermission. When exhausted with these exertions, they settle on walls, trees, or the ground in little parties, and when rested resume their flight. I have seen such flocks as these night after night on the Galle esplanade, and often observed them flying round and round high above the fort before finally moving off for the night to some distant and common roosting-place. When its prey consists of beetles, dragonflies, or other large insects, which it espies from its perch, it is captured after a sometimes prolonged flight, brought back, and killed before being swallowed by being repeatedly struck against whatever object the bird is seated on. This may often be witnessed when the bird is perched on telegraph-wires, which are a very favourite look-out with it. I have seen it dash on to the surface of ponds and rivers, and seize insects which were passing over the water. Mr. Holdsworth has observed it hunting close to the surface of the sea, at a distance of a quarter of a mile from the shore. Jerdon notices its habits of congregating together, and writes that on one occasion he saw an 'immense flock of them, probably many thousands, at Caroor, on the road from Trinchinopoly to the Nilghiris.' They were sallying out from the trees lining the road for half an hour or so, capturing insects, and then returning to them again. As a rule they do not consort in close company, but live in scattered flocks of about half a dozen, and often one or two birds constantly frequent the same locality. The note is difficult to describe. Jerdon not inaptly speaks of it as 'a full mellow rolling whistle.' This Bee-eater retires late to roost, collecting to one spot from many miles round, and forming a large colony, which pass the night in thickly foliaged trees or bushes. On Karativoe Island I discovered one of these roosting-places; the birds were flying over from the mainland some miles distant, and continued to arrive from various points on the opposite coast until it was too dark to distinguish them on the wing. They resorted to the borders of a small back-water beneath the high sand hills of the island, which was lined with mangrove-trees, the thick branches of which afforded them a safe refuge."

Lieut. Kelham, who met with this Bee-eater at Singapore, writes (*l. c.*) that "they arrive there in great numbers towards the end of September, keeping in flocks of from ten to twenty, and frequenting low-lying ground and wet paddy-fields, over which they hawk for insects, at one moment swooping down at a great pace close to the ground, the next rising high into the air and sailing along without a move of their wings; when at rest they are generally to be seen on some conspicuous isolated spot, such as the top of a post or the highest branch of a dead tree. I think I may put it down as migratory; for, on reference to my notes, made daily, I can find no record of its occurrence except during the wet season."

I am indebted to Captain Bingham for the following notes :—" Whenever I have come across this bird I have found it a migrant. At Virgola, on the west coast of India, I met it in numbers in January, but by April not one was to be seen. At Delhi and at Allahabad, in the North-west Provinces, they similarly arrived in November, were scarce in June, and not to be found by July and August. Again in Tenasserim, at Moulmein, and in the interior in the Houndraw valley, they suddenly appeared after the rains, and vanished by the end of the May following. It seems to me more gregarious than the others, and I have observed thirty or forty, not exactly in a flock, but sitting and hawking about over the same patch of paddy-land. It has a fine circling flight, during which it frequently utters its clear rolling whistle."

Like its allies, the present species places its eggs in a hole tunnelled in a river-bank. The nest-hole is said to vary from three to nine feet in depth, and is slightly enlarged, so as to form a sort of chamber at the end. No nest is made; but the eggs, which vary from three to six in number, are usually deposited on the ground in the nest-chamber without any nest-lining; they are rosy pink in tinge when unblown, but when emptied of their contents are pure white and very glossy in texture of shell, and in size measure about 0·95 by 0·81 inch, being somewhat large in proportion. Captain Bingham, in a letter to me respecting the habits of this bird, says :—" At Delhi I found their nest-holes in the banks of sandy nullahs in April, May, and June; at Kaukarit, on the Houndraw, in April, in the banks of the Kaukarit stream, a small feeder of the Houndraw. The tunnel they dig is often more than 7 feet in depth, and the egg-chamber, proportionally larger than that of the smaller species, is, unlike theirs, sometimes lined with a little grass, a few feathers, or the wings of white ants. The eggs vary f om three to five in number, and are of course roundish, pinky white in colour, and glossy."

The specimens figured are those described, and are in my own collection.

In the preparation of the above article I have examined the following specimens :—

E Mus. H. E. Dresser.

a, b. India; *c, ♂*. Secunderabad, 25th November, 1869 (*Marshall*). *d, ♂* ; *e, ♀*. Bela Oudh, 28th April, 1870. *f,g.* Ceylon (*Holdsworth*). *h,♂*. Kaukarit, Tenasserim, 18th June, 1879 (*C. T. Bingham*). *i,♂*. Kaukarit, 2nd May, 1879 (*Bingham*). *k.* Java. *l,* ad.; *m,*juv. Malacca. *n, o.* Negros.

E Mus. Tweeddale.

a. Coorg. *b, c.* N.E. India. *d, e,f.* Candeish. *g.* Deyra Doon. *h,♂*. Hangwella, 23rd March, 1866. *i.* Hangwella, 10th December, 1865. *k.* Hangwella, 2nd January, 1866 (*S. Chapman*). *l, m.* Nicobars. *n, o,p, q, r, s.* Ceylon (*Nevill*). *t.* Thayetmyo. *u.* Karen Hills, November 1874 (*Wardlaw Ramsay*). *v, ♂*. Tonghoo, 4th October, 1875. *w, ♂*. Tonghoo, 20th May, 1875 (*Wardlaw Ramsay*). *x.* Near Bangkok, Siam, 8th December. *y.* Near Bangkok, Siam, 1st December, 1872. *z, ♀*. Monte Alban, February 1877. *aa,♂; bb, ♀*. Valencia, August 1877. *cc, ♂*. Basol, July 1877. *dd.* Basol, October 1877. *ee.* San Mateo, February 1877. *ff,gg.* Malacca, 1873 and 1874. *hh.* Sumatra (*Bock*). *ii.* Sumatra (*Wallace*). *kk.* Java. *ll.* Celebes. *mm.* Celebes. *nn.* Luçon, 7th February, 1872 (*Meyer*).

BLUE CHEEKED BEE EATER
MEROPS PERSICUS

MEROPS PERSICUS.

BLUE-CHEEKED BEE-EATER.

Merops persica, Pallas, Reise Russ. Reichs, ii. Anh. p. 708 (1773) ; Keyserl. & Blasius, Wirbelth. Eur. pp. xxxv
    & 149 (1840) ; Von der Mühle, Orn. Gr. p. 32 (1844) ; C. L. Brehm, Vogelfang, p. 50 (1855) ; Brea, B. of
    Eur. iii. p. 164 (1867) ; C. W. Wyatt, Ibis, 1870, p. 10 ; Schalow, J. f. O. 1877, p. 197.
Merops ægyptius, Forsk. Descr. Animal. Aves, p. 2 (1776) ; Lesson, Traité d'Orn. p. 237 (1831) ; Gray, Gen. of B.
    i. p. 86 (1846) ; Bp. Consp. Gen. Av. i. p. 161 (1850) ; v. Müll. J. f. O. 1855, p. 9 ; Jerdon, B. of Ind. i.
    p. 209 (1862) ; Schlegel, Mus. Pays-Bas, *Merops*, p. 2 (1863) ; Tristram, Ibis, 1864, p. 433 ; Heuglin, Ibis,
    1864, p. 334; Giglioli, Ibis, 1865, p. 52 ; Monteiro, P. Z. S. 1865, p. 96 ; Tristram, Ibis, 1866, p. 83 ; Degland &
    Gerb. Orn. Eur. i. p. 173 (1867) ; Layard, B. of S. Afr. p. 69 (1867); Bocage, Jorn. Ac. Sc. Lisb. ii. p. 134 (1867);
    Gurney, Ibis, 1868, p. 154; Bocage, Jorn. Ac. Sc. Lisb. vii. p. 339 (1870) ; id. op. cit. xiii. p. 66 (1872) ;
    Salvadori, Ucc. d' Italia, p. 44 (1872) ; Hume, Stray Feathers, i. p. 167 (1873) ; Adam, tom. cit. p. 371 (1873) ;
    Savi, Orn. Tosc. 2nd ed. i. p. 238 (1873) ; Hume, Nests & Eggs of Ind. B. p. 103 (1873) ; Sharpe, P. Z. S.
    1874, p. 306; Hume, Str. Feath. iii. pp. 320, 456 (1875) ; Bocage, J. f. O. 1876, p. 407 ; Butler, Stray Feathers,
    vii. p. 181 (1878) ; Wardlaw Ramsay, Ibis, 1879, p. 446.
Le Guêpier savigny, Levaill. Hist. Nat. Guêp. p. 30, pls. 6, 6 bis (1807).
Le Guêpier rousse-gorge, Levaillant, op. cit. p. 52, pl. 16 (1807).
Merops longicauda, Vieill. Nouv. Dict. xiv. p. 15 (1817, ex Levaill.).
Merops superciliosus, Vieill. Nouv. Dict. xiv. p. 20 (1817, nec Linn.) ; Licht. Verz. Doubl. p. 13 (1823) ; Rüpp. Syst.
    Uebers. p. 23 (1845) ; Heugl. Orn. N.O.-Afr. i. p. 197 (1869); Blanford, Geol. & Zool. of Abyss. p. 321
    (1870) ; J. H. Gurney, Ibis, 1871, p. 74 ; Reichenow, J. f. O. 1877, p. 21 ; Bocage, Orn. d'Angola, p. 87 (1881).
Merops savignyi, Audouin, Expl. somm. Pl. Hist. Nat. de l'Egypte (1825, ex Levaill.) ; Crespon, Orn. du Gard,
    p. 291 (1840) ; Malherbe, Faun. Orn. Sic. p. 141 (1843) ; Bp. Consp. Gen. Av. i. p. 161 (1850) ; C. L.
    Brehm, Vogelfang, p. 50 (1855) ; von Müller, J. f. O. 1855, p. 9 ; Hartl. Orn. W.-Afr. p. 39 (1857) ; Layard,
    B. of S. Afr. p. 69 (1867).
Merops savignii (Aud.), Steph. in Shaw's Gen. Zool. xiii. pt. 2, p. 74 (1825) ; Cuvier, Règne Animal, 2nd ed. i.
    p. 442 (1829) ; Swainson, Classif. of B. ii. p. 333 (1837) ; Temm. Man. d'Orn. 2nd ed. iv. p. 649 (1840) ;
    Gray, Gen. of B. i. p. 86 (1846) ; Licht. Nomencl. Av. p. 66 (1854) ; Gurney, Ibis, 1861, p. 132 ; Monteiro,
    Ibis, 1862, p. 334 ; G. F. L. Marshall, Ibis, 1872, p. 203 ; Bocage, J. f. Orn. 1876, p. 407.
Merops persicus, Pall., Keys. & Blas. Wirbelth. Eur. p. 35 (1840) ; Schlegel, Rev. Crit. p. 53 (1844) ; Gray,
    Gen. of B. i. p. 86 (1846) ; Degland, Orn. Eur. i. p. 618 (1849) ; Blyth, Cat. Mus. As. Soc. p. 51 (1849) ;
    Tristram, Ibis, 1859, p. 27 ; Lindermayer, Vög. Griechenl. p. 45 (1860) ; S. S. Allen, Ibis, 1862, p. 359 ;
    Wright, Ibis, 1864, p. 73 ; E. C. Taylor, Ibis, 1867, p. 56 ; Shelley, Ibis, 1871, p. 48 ; Wright, Ibis, 1874,
    p. 237 ; Krüper, Griech. Jahresz. iii. p. 191 (1875) ; Lacroix, Cat. Ois. Pyr. France p. 273 (1873) ; Dresser,
    B. of Europe, v. p. 165 (1877) ; Wardlaw Ramsay, Ibis, 1880, p. 49 ; Scully, Ibis, 1881, pp. 48, 429 ;
    Seebohm, Ibis, 1882, p. 210 ; Tristram, Ibis, 1882, pp. 414, 417 ; Dixon, Ibis, 1882, p. 560.
Blephameroys savignyi (Aud.), Reichenb. Meropinæ, p. 82 (1852) ; Gray, Hand-l. of B. i. p. 99, no. 1206 (1869).
Blephamerops ægyptius (Forsk.), Reichenb. ut suprà (1852) ; Gray, Hand-l. of B. i. p. 99, no. 1205 (1869).
Merops ægyptiaca, Erhard, Fauna der Cycladen, p. 52 (1858).
? *Merops chrysocercus*, Cab. Mus. Hein. i. p. 139 (1859–60); G. F. L. Marshall, Ibis, 1872, p. 203.

Figuræ notabiles.

Levaill. Hist. Nat. Guêp. pls. 6, 6 bis, 16 & 19; Savigny, Hist. Nat. Egypte, pl. 4. fig. 3; Dree, B. of Eur. iii. pl. to p. 162; Shelley, B. of Egypt, pl. vii. fig. 1; Dresser, B. of Eur. v. pl. 206.

Hab. Southern portion of the Western Palæarctic Region, the Ethiopian Region, the south-western portion of the Eastern Palæarctic Region, and it ranges just into the Indian Region.

Ad. suprà lætè psittacino-viridis : fronte albidâ, posticè cærulescente : superciliis et strîâ suboculari cærulescentibus, hâc suprà albo marginatâ : tæniâ per oculum nigrâ : uropygio et supracaudalibus cærulescenti tinctis : mento flavo : gulâ saturatè rufâ : pectore et abdomine cærulescenti-viridibus : subalaribus pallidè ferrugineis : remigibus nigricanti apicatis : rectricibus duabus centralibus elongatis et nigricanti apicatis : rostro nigro : pedibus fuscis : iride coccineâ.

Juv. suprà saturatè cæruleo-viridis, plumis vix pallidiore marginatis : rectricibus centralibus vix elongatis : lineâ frontali nullâ, et superciliis indistinctis : mento pallidè flavo-cervino : gulâ sordidè cinnamomeâ : corpore reliquo subtùs pallidè cæruleo-viridi, abdomine centrali albicante.

Adult male (Egypt).—Forehead and a broad superciliary stripe turquoise-blue, but the forehead at the base of the bill is white; upper parts deep parrot-green, becoming bluish green on the upper tail-coverts, and on the wings and tail tinged with russet; quills tipped with black and on the inner web margined with sooty brown; a broad black band passes from the gape through the eye to the ear, and is narrowly margined below with white, below which there is a broad turquoise-blue stripe; chin dark yellow; throat fox-red; underparts generally deep parrot-green; under wing-coverts and under surface of the wings dull rufous; under surface of the tail blackish grey; bill black; legs dark brown; iris crimson. Total length about 11 inches, culmen 1·8, wing 6·1, tail 6, central rectrices extending 2·1 beyond the lateral ones, tarsus 0·5.

Adult female (Egypt).—Closely resembles the male.

Young (Shiraz, Persia).—Upper parts dark bluish green, much darker than in the adult, the feathers with paler margins; central rectrices scarcely longer than the lateral ones; the frontal blue and white stripes wanting, and the superciliary blue stripe but faintly indicated; the black cheek-stripe margined below with bluish white; chin pale yellowish buff; throat dull light russet; rest of the underparts pale blue-green, becoming very pale on the centre of the abdomen.

Africa appears to be the true home of this Bee-eater, where it is widely distributed; and though it is found in the countries north of the Mediterranean, it is there only met with but rarely, or as an occasional straggler. Crespon states that two examples were killed, in May 1832, near the mouth of the Lez, in the department of Hérault, in Southern France; and according to M. Adrien Lacroix (Cat. Ois. Pyr. Franç. p. 273), one was obtained on the 3rd of May, 1859, near La Nouvelle, in Aude, and he subsequently received one from Cette. In Italy it is of very rare occurrence. The Marquis Durazzo records the occurrence of two which were obtained near Genoa in 1834, one of which is now in the University Museum of that town, and the other in that of Florence; and Professor Giglioli states ('Ibis,' 1881, p. 191) that one was shot at Bari "some years ago." According to Malherbe (Faun. Orn. Sic. p. 141), a female was shot near

Palermo, in Sicily, which he saw; but both Doderlein and Benoit doubt this occurrence. Schembri states that one was killed in Malta in September 1840; but Mr. C. A. Wright remarks that all trace of this specimen has been lost. This latter gentleman, however, possesses a specimen which he says was "killed at the end of May 1871, at the Inquisitor's Palace, Malta, by F. Camilleri, barber of the Central Hospital, out of a flock; but whether of the same or of the common kind (*M. apiaster*) he could not say. He was first attracted by its note, which was different from any he had heard before.

It is included amongst the birds of Greece by both Lindermayer and Von der Mühle, who say that they have on several occasions found it exposed for sale in the Athens market amongst common Bee-eaters; and Erhard records it from the Cyclades as a "summer visitant," under which heading he includes the birds which breed in those islands. Dr. Krüper also states that a small flock of these Bee-eaters was seen in Greece on the 19th of April, 1874, out of which four specimens were obtained.

In Southern Russia it is of very rare occurrence. Professor von Nordmann observed it twice near Odessa, and according to Hencke (Ibis, 1883, p. 210) six examples were once obtained at the mouth of the Volga late in May. Canon Tristram met with this Bee-eater near Berejik, in Armenia, and writes (Ibis, 1882, p. 414), "I was delighted to find here, for the first time in any numbers, a colony of the Persian Bee-eater (*Merops persicus*), not so numerous as *M. apiaster*, but still plentiful. The habits of the two species are markedly different when seen together. *M. persicus* is by no means shy, and perches much more frequently than the other, settling on low trees, and frequently on the top of a thistle-tuft." Canon Tristram shot one in the Jordan Valley in 1858, and Mr. Cochrane, who accompanied him on his second journey, saw a flock near Hebron. I may also here state that it has been obtained at Beyrout, in Syria.

In North-east Africa the present species is very common. Captain Shelley says that it is "the most abundant of the Bee-eaters in April. It arrives in the country about a fortnight earlier in the spring than *Merops apiaster*." Mr. E. C. Taylor saw the first at Benisouef on the 26th of March, after which time they became plentiful. They were, he adds, very tame and much given to perching on telegraph-wires. Von Heuglin writes (*l. c.*):—"In the latter days of March, and usually before *M. apiaster* arrives, small flocks of this species appear on passage in Lower Egypt, and frequent fields, gardens, and fig-plantations, on the edge of the desert, the dunes, or in meadows, and usually leave after a sojourn of a few days, to return again in hundreds in June and July, when they often collect together, several hundred in a flock, and are seen chiefly in the olive-gardens, and on tamarisks and acacias along the canals. In the morning they remain where they have roosted, utter their call-note in a low tone, and about nine o'clock collect in flocks, and spread over the fields and in the villages, uttering loud cries. Their flight is Swallow-like, but irregular, and one or two leave the flock and circle round catching insects, which are devoured either on the wing or when seated on a branch at the top of a tree. During the heat in the middle of the day they rest for a time; and I never saw a Bee-eater go to drink. In the evening they collect together and, uttering their note noisily, go to roost. In the summer they are very fat, and numbers are killed and eaten by the Italian and Greek gunners. Late in August one meets flocks of this Bee-eater on passage in Nubia, East Sudan, and Abyssinia; but they do not winter here, but migrate further in a southerly or south-westerly direction. On the 17th of October, 1857, I found the *Avicennia*-thickets on some of the islands on the Somali coast covered with Bee-eaters and Rollers, which evidently came there after the flocks of locusts. Brehm surmised correctly that this species breeds in May, in Central and Lower Egypt; for I

K

shot a female at New Dongolah on the 19th May which had a fully formed egg in the ovary. Allen found a colony breeding at Damietta in April; and I also found one on some half-desert pasture-land at Dachschur in the same month. But Brehm is wrong in saying that all Bee-eaters (and he probably means *M. apiaster* in particular) migrate solely in company with *M. persicus*; for one finds separate flocks of the different species in the same locality, but I never saw them intermixed. Hartmann states that he observed *Merops persicus* in January near Golosaneh in Egypt, whereas I never saw it between September and March in North-east Africa. In the autumn the plumage fades greatly and loses the rich green sheen; and the moult probably takes place in January and February."

Mr. Jesse records the Bee-eater from Abyssinia; but Mr. Blanford did not meet with it, though he obtained one, shot at Adigrat by Capt. Newport.

In North-western Africa it is far less common than on the north-eastern side of the continent; and though it occurs in Algeria, it is far from common there. Mr. J. H. Gurney says (Ibis, 1871, p. 75):—"On the 21st of April I saw an Egyptian Bee-eater in one of the cemeteries at Gardaia, which proved to be of this species. I afterwards came upon a flock of them lying upon the large stones which are scattered about wherever there are no gardens. I saw them also on walls, and on the fence-work upon the town-wall; and returning I found the cemetery, where I had seen the first solitary bird, occupied by about a dozen. They were perfectly tame; and I thought I had never seen a more interesting sight than these sparkling birds as, one after another, they rose into the air to hawk for insects, and, returning, perched upon a tombstone within a few yards, perhaps, of where I was standing. They have only one note; it is loud and rather harsh, like the Common Bee-eater's. Their flight is slower, but even more gliding, with the wings very much raised, except when the birds are high in air, when they appear to be more depressed. They almost lie upon stones and walls, as if unable to sit upright on account of their long tails and short legs." On the west side of the African continent the Blue-cheeked Bee-eater is found as far south as the Cape Colony. Swainson records it from Senegambia, and there are specimens from Senegal in the Leyden Museum. Verreaux records it from Casamanze, Bissao, and the Gaboon, Perrein from Malimbe, and Monteiro from Benguela and Angola. This last explorer says (*l. c.*) that in Angola it was " generally seen on the tops of trees, from which it darts out and sweeps slowly in the air, in the manner of a Swallow, returning to rest on the tree, where it utters a very peculiar and mournful cry; their stomachs contained remains of insects." There are examples in the Lisbon Museum from Loanda and Rio Quilo in Angola; and Professor Barboza du Bocage writes (*l. c.*) that " it appears, though not commonly, in the southern portions of the Portuguese settlements of Angola, chiefly towards the interior, as M. Anchieta does not seem to have observed it in the vast tract he traversed from Capangombe to Cunene. Andersson only observed it once on the Okovango river. It disappears or becomes very rare in localities where *Merops apiaster* is common." I possess specimens from Bissao and from Ondonga, the latter obtained by Mr. Andersson, who observed it near the Okavango river; and in the last collection he sent over there were several specimens obtained in Ondonga, in November 1866. Mr. E. L. Layard, who records this Bee-eater from the Cape Colony, writes that it was " found in Natal by Mr. Ayres. A single specimen was also forwarded to the Museum by T. B. Bayley, Esq., of Wynberg, having been shot by that gentleman on the Cape Flats. Mr. Dumbleton, of Wynberg, assures me that these birds periodically visit a circumscribed portion of the Cape Flats in considerable numbers. On the 15th February, 1866, a specimen shot near Kuils river was sent to the Museum by Mr. Bishop."

As above stated, Mr. Ayres met with this species in Natal; and Mr. T. E. Buckley, who met with it in the Matabili country, writes (Ibis, 1874, p. 363), "This species was only observed on one occasion, when it appeared to be migrating. On that day I saw several large flocks hawking about after flies and occasionally settling on the small bushes."

Mr. Ayres says (Ibis, 1861, p. 132):—"These birds take their food on the wing, and their flight somewhat resembles that of the Swallows; they frequently alight on the trees and bushes to rest; during flight they utter a harsh grating note. I believe they only inhabit the coast lands, and are migratory, appearing only in the summer months."

On the east coast of Africa, where *Merops superciliosus* is also found, it is somewhat difficult to define the range of the two species; but *M. persicus* appears to occur in the Zambesi country, as there is a specimen in the British Museum obtained there by Bradshaw, and Captain Shelley possesses two examples obtained at Mahalaka by the same collector.

In Asia this Bee-eater ranges into Western India. Mr. C. W. Wyatt obtained it at El Noweyba, on the peninsula of Sinai. Heuglin records it from Arabia, Captain Jones from Mesopotamia, and it is said to be common at Orenberg and on the shores of the Caspian and Aral and in Turkestan, in which last country Dr. Severtzoff states that it breeds. Mr. Blanford says he " found *M. persicus* in great abundance in the country north-west of Bampur, in Baluchistan, and in Narmashir, the Persian district traversed on the road from Bampur to Bam, in the second and third weeks of April 1872. The birds were evidently migrating, and all which were shot were in superb plumage. Hume remarks that large numbers are seen in Sind at particular seasons, probably in the same manner, when migrating, and the bird has been observed as far east as Aligarh. On the Persian highlands I seldom saw this species, *M. apiaster* being much more abundant; but a few miles from Tehrán, on the 22nd August, I came upon a large scattered flock of *Merops persicus*, chiefly consisting of young birds. The place was a somewhat barren plain, with a few scattered shrubs and herbaceous plants; and the birds settled on the ground, occasionally flying up to pursue insects. They may have been migrating, or preparing to migrate. De Filippi obtained specimens at Miána and Nikbeg, between Kazvin and Tabriz; and Ménétriés saw it on the banks of the Kur, in the Transcaucasian provinces of Russia, a little north of the Persian frontier." Capt. Wardlaw Ramsay observed large flocks on migration in Afghanistan, late in April and early in May. Captain Butler (Str. Feath. iii. p. 457) observed it " on several occasions near Deesa," and remarks that it has a fine wild note, which it utters on the wing, and which much resembles the note of *M. apiaster*; and Mr. Hume remarks that he has specimens or records of this bird, but only as a summer visitant, from both Northern and Southern Sind, Cutch, Kattinwar, Jodhpoor, and indeed the whole of Rajpootana. Mr. Scully observed it in Gilgit, between the 20th and 28th November, 1879, when several flocks passed the valley on migration southwards. Captain Marshall obtained it in the Aligurh and Myapoorie districts in the North-west Provinces of India. Mr. R. M. Adam says (Str. Feath. i. p. 371) that "it is rarely seen about Sambhur, but about the tree- and scrub-jungle at Mata Pahar and the Marot hills it is very common. In the Marot hills the natives showed me the holes in which it breeds, about the beginning of the rains, and Capt. Bingham found it breeding in Upper India, at the Sultanpoor salt-works near Delhi, where hundreds may be seen, but he did not observe a single *M. philippensis*." I am indebted to the last gentleman for the following notes:—"I know very little about this bird, having met with it only at Delhi, where, and to the country to the south of it, it comes in, in great numbers, in the beginning of the hot weather to breed. My observations led me to believe that all the young were out of the nests by July, when either all the old birds migrated, leaving only the young behind, or they moulted, and were then undistinguishable from the latter. However this may be, it is certain that after the 15th July I never procured a full-plumaged bird of this species until the hot weather following, while specimens, whether actually young birds of the year or not, I cannot say, were to be had every day throughout the autumn, in seemingly immature plumage." With regard to the habits of this Bee-eater, Captain Bingham writes:—"The habits of this species closely resemble those of *M. philippinus*, for which it can be easily mistaken. Like the latter, it is more or less gregarious and given to circling round in lofty flights, uttering a whistle precisely similar to the whistle of that species. I have taken several nests; these resemble those of *M. philippinus* closely, but are, as well as I can remember (I cannot find the notes I took on the spot about them), unlined. The eggs are, as is the unvarying rule in the Meropidæ, white, almost globular, and highly glossed. In size they may be a trifle larger than those of *M. philippinus*."

Captain Shelley says (*l. c.*) that "the Blue-cheeked Bee-eater resembles *Merops apiaster* in size, habits, and cry; yet the two species are never found in one flock. During the day they may generally be met with perched upon the telegraph-wires, or feeding among the herds of cattle. I once observed them, towards evening, alight in such immense numbers upon a sand-bank, that they made it look almost as green as meadow-land; they appear, however, generally to roost at night in the sont trees."

The late Mr. S. Stafford Allen, who found the present species breeding on the banks of the Nile, says (Ibis, 1862, p. 359):—"These birds mostly fly in flocks of twenty or thirty, though sometimes in much greater numbers. Whilst on their way in the daytime they keep at a considerable height, and sail about like Swallows, though not so rapidly, descending at night to roost in trees. They have a sharp twittering cry, which is often distinctly audible when the birds are almost out of sight. The Arab name of 'Dar-doon' is applied to both species .... Whilst returning from an ornithological excursion down the Nile to Damietta on the 21st of April (1862), our attention was attracted by a large flock of *M. persicus* hovering over one particular spot, where others of their number were settled on the ground. On a closer examination, a large number of holes were seen in a piece of ground between the river and a field of young wheat, which very slightly shelved down towards the water, in and out of which holes Bee-eaters were constantly passing. After digging out a passage of nearly 4 feet in length, which went in at an angle of 10° or 15°, we found a slightly enlarged chamber, which formed the nest. The bottom of this chamber was covered with the remains of dragon-flies, &c. (mostly wings), upon which the eggs were deposited. These were of a pure white, nearly round, and about 10 lines in length. The greatest number found in any one nest was three; but the birds had evidently only just begun to lay (many of the holes being unfinished), so that we were unable to ascertain what is the usual number deposited. More than forty holes were opened, but only eleven eggs obtained. In the vicinity of every hole were numbers of pellets, formed of the wings and other indigestible parts of dragon-flies, butterflies, beetles, &c., which had been cast up by the Bee-eaters in the same manner as Hawks and Owls."

I possess several eggs of this Bee-eater, which do not in any way differ from those of *Merops apiaster*.

The specimens figured and described are an adult male from Egypt and a young female from Persia, both of which are in my own collection.

In the preparation of the above article I have examined the following specimens :—

E Mus. H. E. Dresser.

a, ♂; b, ♀. Egypt (*Captain Shelley*). *c, ♂*. Benisouef, Egypt, 28th March, 1868 (*Capt. Shelley*). *d, ♂*. Egypt (*Rogers*). *e, ♂* ad. Regan, Narmashir, S.E. Persia, 18th April, 1872 (*W. T. Blanford*). *f, ♀* juv. Shiraz, Persia, summer of 1870 (*W. T. Blanford*). *g*. Bissao, Africa (*Verreaux*). *h*. River Gambia (*Whitely*). *i*, juv. Ondonga, Ovampo, 28th November, 1866 (*Andersson*). *k, ♂* ad. Djidili Kuduk, Turkestan, 13th June (*Severtzoff*). *l, ♂* ad. Nukus, on the Oxus, 20th August (*Severtzoff*). *m, ♀* ad. Petroalexandrowsk, on the Oxus, 6th September (*Severtzoff*). *n, ♂*. Lankoran, 27th May. *o, ♀*. Lankoran, 17th April.

E Mus. Tweeddale.

a. Egypt. *b, ♀*. Koomaylee, 22nd March, 1868 (*W. Jesse*). *c*. Zambesi. *d*. Ondongo, Ovampo, 22nd December, 1866 (*Andersson*). *e*. Gambia. *f, ♀*. Sultanpur, 8th June, 1876.

E Mus. G. E. Shelley.

a, ♀. Egypt, 25th March, 1870. *b*. Egypt, March 1871. *c, ♂*. Egypt, 29th March, 1871. *d, e, ♂*. Egypt, 1st April, 1871 (*G. E. Shelley*). *f, ♂*. Damietta, Egypt, April 1878 (*Filipponi*). *g*. Durban. *h*, juv. Durban. *i, k*. Mahalaka (*Bradshaw*).

E Mus. Brit.

a, b. North Africa (*Gould coll.*). *c*. Djeddah. *d*. Dongola (*Schaufuss*). *e*. Zambesi (*Bradshaw*). *f*. River Gambia, March 1864 (*Whitely*). *g, ♂*. Ondonga, 20th November, 1866 (*Andersson*). *h*. Benguela (*Monteiro*). *i*. Benguela, 1870. *k*, juv. Junction of Mooi and Vaal rivers (*Barratt*). *l*. Bissao (*Verreaux*). *m, ♂*. Henrico Vley. *n*. Tette (*Livingstone*). *o*. Euphrates expedition. *p, ♂*. W. of Bampur, Baluchistan, 18th April, 1872 (*Blanford*). *q, ♂* juv. Shiraz, Persia, 1870 (*Blanford*). *r*. Mesopotamia (*Capt. Felix Jones*). *s*. Candahar (*Griffith*). *t*. Candahar (*Gould coll.*).

WHITE CHEEKED BEE EATER

MEROPS SUPERCILIOSUS

MEROPS SUPERCILIOSUS.

MADAGASCAR BEE-EATER.

Apiaster madagascariensis, Briss. Orn. iv. p. 546, pl. 42. fig. 1 (1760).

Merops superciliosus, Linn. Syst. Nat. i. p. 283 (1766) ; Lath. Ind. Orn. p. 271 (1790) ; Bechstein in Lath. Uebers. der Vög. iv. p. 160 (1811) ; Shaw, Gen. Zool. viii. pt. 1, p. 164 (1811) ; Vieill. Nouv. Dict. xiv. p. 9 (1816) ; Cuv. Règn. Animal, 1st ed. p. 415 (1817) ; Vieill. Tabl. Encycl. i. p. 278 (1823) ; Licht. Verz. Doubl. p. 13 (1823) ; Steph. in Shaw's Gen. Zool. xiii. pt. 2, p. 73 (1825) ; Cuvier, Règn. Animal, 2nd ed. i. p. 442 (1829) ; id. 3rd ed. Ois. p. 199 (1836) ; Temm. Tabl. Méthod. p. 54 (1838) ; id. Man. d'Orn. iv. p. 649 (1840) ; Gray, Gen. of B. i. p. 86 (1846) ; Bp. Consp. Gen. Av. i. p. 161 (1850) ; Licht. Nomencl. Av. p. 66 (1854, partim) ; Von Müller, J. f. O. 1855, p. 9; Kollar, Sitz. Ak. Wiss. Wien, 1858, p. 342 ; Cab. Mus. Hein. ii. p. 140 (1859) ; Hartlaub, J. f. O. 1860, p. 87 ; id. Orn. Beitr. Faun. Madag. p. 32 (1861) ; id. J. f. O. 1861, p. 106 ; Roch & Newton, Ibis, 1862, p. 272 ; Pollen, Nederl. Tijdsch. v. d. Dierk. 1863, p. 311 ; E. Newton, Ibis, 1863, p. 341 ; Sclater, Ibis, 1864, p. 290 ; A. Newton, P. Z. S. 1865, p. 834 ; Verreaux, Voy. à Madag. de Vinson, Ann. B, p. 1 (1865) ; Schlegel, P. Z. S. 1866, p. 421 ; Finsch, J. f. O. 1867, pp. 239, 245 ; Grandidier, Rev. et Mag. Zool. 1867, p. 355 ; Sperling, Ibis, 1868, p. 288 ; Schlegel & Pollen, Faun. de Madag. ii. p. 60 (1868) ; Finsch & Hartlaub, Vög. Ost-Afr. p. 181 (1870, partim) ; Sharpe, P. Z. S. 1870, p. 397 ; id. Cat. of Afr. B. p. 3 (1871) ; Andersson, B. of Damara Land, p. 61 (1872) ; Reichenow, J. f. O. 1875, p. 18 ; Bartlett, P. Z. S. 1875, p. 65 ; Sharpe, in Layard's B. of S. Afr. p. 97 (1875, partim) ; E. Newton, P. Z. S. 1877, p. 297 ; Hartl. Vög. Madag. p. 81 (1877) ; Pollen, Relat. du Vog. F. de Madag. i. pp. 96, 123, 133 (1877) ; Ayres, Ibis, 1878, p. 285 ; Fischer & Reichenow, J. f. O. 1878, p. 256 ; Nicholson, P. Z. S. 1878, p. 355 ; Sharpe, in Oates's Matabele Land, p. 301 (1881).

Le Guêpier de Madagascar, D'Aubenton, Pl. Enl. no. 259.

Le Patirich, Montb. Hist. Nat. Ois. vi. p. 495 (1779).

The Supercilious Bee-eater, Lath. Gen. Synops. i. p. 673 (1783).

Le Guêpier rousse tête ou le Guêpier bonelli, Levaill. Hist. Nat. Guêp. p. 57, pl. 19 (1807).

Merops ruficapillus, Vieill. Tabl. Encycl. et Méth. p. 391 (1820).

Merops ægyptius, var., Smith, S. Afr. Quart. Journ. ii. p. 320 (1833).

Merops vaillantii, Bp. Consp. Gen. Av. i. p. 161 (1850) ; Von Müller, J. f. O. 1855, p. 9.

Merops ægyptius, Reichenb. Meropinæ, p. 64 (1852, nec Forsk.).

Blepharomerops ægyptius, Reichenb. op. cit. p. 82 (1852, nec Forsk.).

Blepharomerops superciliosus (Linn.), Reichenb. ut suprà (1852) ; Bp. Vol. Anisod. ii. p. 318 (1854) ; Gray, Hand-l. of B. i. p. 99, no. 1204 (1869).

Merops madagascariensis typicus, Milne-Edw. & Grandidier, Hist. Nat. Ois. Madag. p. 262, pls. xc., xcii. (1881).

Saint-esprit, Sicirici-rico, Kirio-kirio, in Madagascar.

Figuræ notabiles.

D'Aubenton, Pl. Enl. 259 ; Levuillant, Hist. Nat. Guêp. pl. 19 ; Reichenbach, pl. cccexliii. B. figs. 3545, 3546.

M

Hab. Madagascar and adjacent islands; south-west and south-eastern portions of the mainland of Africa.

Ad. *M. persico* similis, sed suprà saturatior: pileo viridi-fusco: lineâ frontali et superciliis albis vix viridi tinctis, striâ sub vittâ nigrâ per oculos ductâ albâ: mento summo albo vix flavo lavato: corpore subtùs sicut in *M. persico* colorato, sed pallidiore: rostro, pedibus et iride sicut in *M. persico* coloratis.

Adult male (Madagascar).—Upper parts generally as in *Merops persicus*, but rather darker; crown much darker, being of a coppery brownish-green tinge; a narrow frontal line continued to a superciliary line over the eye white, with a faint greenish tinge; the stripe below the black patch through the eye and the upper part of the chin also white, the latter tinged with yellow; the chestnut-red on the throat slightly paler than in *M. persicus* and broader, forming the white stripes below the eye; underparts as in *M. persicus*, but rather paler; bill, legs, and iris as in *M. persicus*. Total length about 12 inches, culmen 1·85, wing 5·4, tail 6·0 (central rectrices elongated and attenuated, extending 2·6 beyond the lateral ones), tarsus 0·55.

This species (which appears to me to be perfectly distinct from *Merops persicus*, although it has by so many ornithologists been united with that species) inhabits Madagascar, Anjuan, and the south-western and eastern parts of the mainland of Africa, and is even found in South-eastern and Western Africa. Hartlaub even states that it has been obtained as far north as Gaboon, and writes (Faun. Madag. p. 32) that he compared an example obtained by Gujon and found no appreciable difference between it and specimens from Madagascar. Reichenow records it from the Loango coast, and Andersson from Damara Land; but a specimen obtained by the latter explorer at Ondonga, which I have examined, I should certainly refer to *Merops persicus*. One, however, obtained by Sala in Angola, as well as one obtained by Monteiro in the same country, both of which are in the British Museum, agree closely with Madagascar examples. Mr. Sharpe considers that this species is not separable from *Merops persicus*, and writes (in his edition of Layard's B. of S. Afr. p. 97) that "notwithstanding the difference in the shades of blue and green which are to be found in a series of skins of this Bee-eater, we believe that but one species is represented; the brown head which is sometimes seen, more especially in Madagascar birds, is often to be noticed in specimens from other parts of Africa, and these brown-headed individuals occur along with green-headed specimens, so that they are nothing but immature birds." In this view, however, I cannot in the least concur; for though it is true that the brown-headed and white-cheeked bird does rarely occur in West and South Africa, together with the blue-cheeked green-headed Bee-eater, yet in East Africa the former appears to be the predominant species, and in Madagascar and the adjacent islands it alone occurs, there being no trace of the existence of the Blue-cheeked Bee-eater eastward of the mainland of Africa.

Mr. Sharpe further states (*ut suprà*) that Señor Anchieta obtained *Merops superciliosus* on the Rio Coroca, in Mossamedes, and that it has also been sent from Benguela by Señor Furtado d'Antas. It is also stated to occur in the Cape Colony, and I possess a specimen from the Transvaal. Mr. Sharpe also records it as being found in Matabele Land. Captain Sperling obtained it in the Mozambique Channel; and Messrs. Fischer and Reichenow record it from Zanzibar, and Mr. Nicholson (P. Z. S. 1878, p. 355) from Dar-es-Salaam, opposite Zanzibar. Dr. Kirk has obtained it on the east mainland of Africa, and I have examined specimens sent by hiu from Melinda and the Pangani river, and he records it from the Zambesi. It inhabits the Comoro Islands. Dr. Kirk says (Ibis, 1864, p. 299) that it was seen on the sugar-plantations near Oane, on the island of Mohilla; and I have examined specimens from Anjuan which agree very closely with Madagascar birds, differing but slightly in having the crown rather less brown and the central rectrices less elongated.

In Madagascar this Bee-eater is said to be very generally distributed. Mr. E. Newton says (Ibis, 1863, p. 341) that it was "to be seen hawking about the Hivondrona river almost daily. On the Fangandrafrah, a tributary of the Hivondrona, I dug out a female Bee-eater from a hole in the bank of the river, about three feet in length; the nest was not yet made, and the bird's beak was covered with soil, showing that she was still working at the excavation. All the specimens, of both sexes, that we obtained were bare of feathers on their breasts and thighs, as if incubating."

Messrs. Pollen and Van Dam write (Faune de Madag. ii. p. 60) that they met with this bird "on the islands of Mayotte, Nossi-bé, Nossi-falie, and in Madagascar, where it is common in places between the promontory of Ambaton, the plains of Syrangene, and along the high sand-banks skirting the Ambassuana river. It is very generally known under the name of Saint-Esprit. It affects the plains on the borders of the forests or the banks of the rivers, and is almost always to be seen perched on the branches of a dead tree in the plains or clearings or on the banks of the rivers or shores of the lakes. These birds quit their perch every minute to hunt after their prey or fly in circles round the tree on which they were perched, uttering in a soft voice the cry *cirio*, *cirio*, and then return to their former resting-place. They sit very upright, but this does not prevent their dashing off and seizing on the wing with great dexterity the insects that pass before them. They are by no means shy, and allow themselves to be approached without showing fear. On being fired at they return almost immediately to their former perch, and do this until shot. These birds live in pairs and but seldom singly, but are sometimes seen in parties of from six to twelve individuals. When nesting they congregate in colonies. When ascending the Ambassuana we saw, about halfway up the beautiful river, in a sort of high rampart of sand, a number of holes surrounded by Bee-eaters, which, flying round incessantly, uttered loud cries. I told my Antancar servants to go ashore and examine these holes; but this they refused to do, because they were afraid of sinking in the mud which was at the foot of the rampart. Curiosity and an earnest desire to ascertain if these holes contained nests urged me to be the first on shore, and I went, sinking up to my knees in the mud. As soon as my Antancars and my Bourbon creole, Eugène, saw me ashore, they hurried after me; but finding it impossible to reach the holes, I had one of my young Antancars tied to a strong rope and let down, and with a wand, to the end of which we had fastened a fork, he examined the holes. Unfortunately all were empty, and the dung at the entrance showed that the young birds had flown. Meanwhile the Bee-eaters came from all sides in large numbers and flew round us uttering lamentable cries, and approached quite close to the young Antancar. These round holes are dug to a depth of about a metre, with an opening about large enough to admit a woman's hand, and the chamber at the end is lined with straws and feathers. The eggs, which are deposited about the middle of October, are almost always two in number, are small for the size of the bird, very fragile, and pure white. The moult takes place in the months of April and May—at least all that we killed in these months were in moult, the long central rectrices being wanting, and they had lost the bright tints of their plumage. The Antancars and Sakalaves call this bird by the name of *Sicirici-rico*, from its cry."

M. Grandidier writes (Hist. Nat. Ois. Madag. p. 262) :—"These Bee-eaters are common on all the coasts of Madagascar, where they inhabit the bush-covered plains, the banks of water-courses, and the openings of forests. They do not occur on the bare mountains in the centre of the island, though one of our party saw one in the great valley of Ampatrana, where the Mangoko runs to the south of the fort of Modongy. Usually they are seen perched on a dead branch of an isolated tree, often on the points of the palisades surrounding the cattle-pens and some of the village houses. When they perceive an insect they dart on it and soon returning take up their old perch, and sometimes they skim along the water in search of their prey. From time to time they rise into the air and circle about like Swallows. Otherwise they resemble, in flight, habits, and cry, the Common Bee-eater, and like them they feed exclusively on insects, which they take on the wing; and they nest in holes about a metre deep, which they bore with their long bill in the argillaceous or sandy banks of rivers, and which they line with dry herbs and feathers. Their eggs, usually two in number, are oval and pure white, measuring 25 by 22 mm. In the nesting-season they live in flocks. They collect in large numbers to roost on the same tree. Their moult appears to take place early in the dry season. They are not wild and are easily shot."

From the particulars respecting the habits of this species which I cite above, it will be seen that it does not differ from its close ally *Merops persicus*, either in its general habits or mode of nidification; and, like that bird, it deposits pure white, roundish eggs, on the ground, at the end of a hole which it tunnels in a bank, usually in the vicinity of water.

The specimens figured and described are in my own collection.

In the preparation of the above article I have examined the following specimens :—

E Mus. H. E. Dresser.

a. Madagascar.   *b.* Anjuan, 1879 (*Bewsher*).   *c.* Transvaal (*Ayres*).

E Mus. Brit.

a. Madagascar.   *b.* Madagascar (*Verreaux*).   *c.* Madagascar (*Crossley*).   *d.* Angola (*Sala*).   *e.* Angola (*Monteiro*).

E Mus. Tweeddale.

a. Madagascar (*Plant*).   *b, ♂.* Mayotte, 6th January, 1864 (*Pollen & Van Dam*).

E Mus. G. E. Shelley.

a. Melinda (*Kirk*).   *b, c, d.* Pangani River (*Kirk*).   *e.* Grand Comoro (*Kirk*).   *f.* Dar-es-Salaam (*E. C. Buxton*).   *g.* Madagascar (*Crossley*).   *h.* Anjuan Island (*Kirk*).

E. Mus. A. & E. Newton.

a, ♂; b, ♀. East coast of Madagascar (*E. Newton*).   *c, ♂; d, ♀.* Anjuan, 1876 (*Bewsher*).

COMMON BEE EATER
MEROPS APIASTER

MEROPS APIASTER.

COMMON BEE-EATER.

Apiaster, Briss. Orn. iv. p. 532 (1760).

Apiaster icterocephalos, Briss. tom. cit. p. 537 (1760).

Merops apiaster, Linn. Syst. Nat. i. p. 182 (1766); Scopoli, Ann. I. Hist. Nat. p. 54 (1769); Gmel. Syst. Nat. i.
p. 460 (1788); Lath. Ind. Orn. i. p. 269 (1790); Licht. Cat. rer. nat. rariss. p. 21 (1793); Wolf, Taschenb.
deutsch. Vögelk. i. p. 132 (1810); Steph. Gen. Zool. viii. pt. 1, p. 152 (1811); Temm. Man. d'Orn. p. 260
(1815); Vieill. Nouv. Dict. xiv. p. 11 (1816); id. op. cit. 2nd ed. i. p. 420 (1820); Naumann, Vög. Deutschl.
v. p. 462, pl. 143 (1826); Lesson, Traité d'Orn. p. 237 (1831); C. L. Brehm, Vög. Deutschl. p. 147 (1831);
Smith, S. Afr. Quart. Journ. 2nd ser. part ii. p. 319 (1834); Swainson, Classif. of B. ii. p. 333 (1837);
Dickson & Ross, P. Z. S. 1839, p. 119; Keys. & Blas. Wirbelth. Eur. p. 35 (1840); Vigors, P. Z. S. 1841, p. 6;
Von der Mühle, Orn. Griech. p. 32 (1844); Schlegel, Rev. Crit. p. 52 (1844); Rüpp. Syst. Uebers. p. 23
(1845); Gray, Gen. of B. i. p. 86 (1846); Degland, Orn. Eur. p. 616 (1849); Blyth, Cat. Mus. As. Soc. Beng.
p. 51 (1849); Bp. Consp. Gen. Av. i. p. 160 (1850); Vern. Harcourt, P. Z. S. 1851, p. 145; Lichtenstein,
Nomencl. Av. p. 66 (1854); Von Müll. J. f. O. 1855, p. 9; Sundevall, Sv. Fogl. p. 160, pl. 70. fig. 1 (1856);
Bolle, J. f. O. 1857, p. 324; Hartlaub, Orn. W.-Afr. p. 38 (1857); Leith Adams, P. Z. S. 1858, p. 474;
id. P. Z. S. 1859, p. 174; Salvin, Ibis, 1859, p. 303; Tristram, Ibis, 1859, p. 435; Cab. Mus. Hein. ii.
p. 141 (1859); Lilford, Ibis, 1860, p. 235; Lindermayer, Vög. Griechenl. p. 44 (1860); Jerd. B. of Ind. i.
p. 210 (1862); Blasius, Ibis, 1862, p. 65; Tristram, Ibis, 1862, p. 278; Homeyer, J. f. O. 1862, p. 255;
Chambers, Ibis, 1863, p. 476; Altum, J. f. O. 1863, p. 113; Homeyer, J. f. O. 1863, p. 263; Schlegel, Mus.
Pays-Bas, *Merops*, p. 4 (1863); Schwaitzer, J. f. O. 1864, p. 315; Heuglin, J. f. O. 1864, p. 334; Tristram,
P. Z. S. 1864, p. 433; Wright, Ibis, 1864, p. 73; Tristram, Ibis, 1866, p. 83; Schneider, J. f. O. 1867, p. 233;
Layard, B. of S. Africa, p. 68 (1867); Chambers & Gerbe, Orn. Eur. i. p. 172 (1867); E. C. Taylor, Ibis, 1867,
p. 56; W. T. H. Chambers, Ibis, 1867, p. 104; C. F. Tyrwhitt Drake, Ibis, 1867, p. 425; Meves, Œfv. K. Vet.-
Akad. Förh. 1868, p. 264; A. C. Smith, Ibis, 1868, p. 449; Gray, Hand-l. of B. i. p. 99, no. 1201 (1869);
Meves, J. f. O. 1869, p. 391; Heuglin, Orn. N.O.-Afr. i. p. 196 (1869); Blanford, Geol. & Zool. Abyss.
p. 320 (1870); Bocage, J. Ac. Sc. Lisb. viii. p. 340 (1870); Finsch, Tr. Zool. Soc. vii. p. 223 (1870); Wyatt,
Ibis, 1870, p. 12; Elwes & Buckley, Ibis, 1870, p. 189; Shelley, Ibis, 1871, p. 48; Saunders, Ibis, 1871,
p. 67; J. H. Gurney, jun., Ibis, 1871, p. 74; Fritsch, J. f. O. 1871, p. 188; Rey, J. f. O. 1872, p. 143;
Shelley, B. of Egypt, p. 169 (1872); Salvadori, Ucc. d'Ital. p. 44 (1872); Jordon, Ibis, 1872, p. 3; G. F. L.
Marshall, Ibis, 1872, p. 203; G. C. Taylor, Ibis, 1872, p. 230; A. B. Brooke, Ibis, 1873, p. 236; Hume,
Nests & Eggs of Ind. B. p. 103 (1873); Savi, Orn. Tosc. ed. i. p. 325 (1873); Bocage, J. Sc. Ac. Lisb. xvii.
p. 35 (1874); Wright, Ibis, 1874, p. 237; Buckley, Ibis, 1874, p. 363; Shelley, Ibis, 1875, p. 69; Danford &
Harvie Brown, Ibis, 1875, p. 300; Blanf. E. Pers. ii. p. 122 (1876); Bocage, J. f. O. 1876, p. 407; Urephcen,
J. f. O. 1878, p. 133; Rud. v. Austria, J. f. O. 1879, pp. 50, 115; Wardlaw Ramsay, Ibis, 1879, p. 446; id.
Ibis, 1880, p. 49; Scully, Ibis, 1881, p. 48; Bocage, Orn. d'Angola, p. 86 (1881); Sharpe, in Oates's Matabele
Land, p. 301 (1881).

Le Guêpier, Montb. Hist. Nat. Ois. vi. p. 480, pl. xxiii. (1779).

Le Grand Guêpier vert et bleu à gorge jaune, Montb. tom. cit. p. 502 (1779).

Le Guêpier à tête jaune, Montb. tom. cit. p. 510 (1779).

Merops congener, Gmel. Syst. Nat. i. p. 461 (1788) ; Shaw, Gen. Zool. viii. pt. 1, p. 155 (1811).

Merops chrysocephalus, Gmel. tom. cit. p. 463 (1788) ; Shaw, Gen. Zool. viii. pt. 1, p. 176 (1811) ; Vieill. Nouv. Dict. xiv. p. 25 (1816).

Le Guêpier vulgaire, Levaill. Hist. Nat. Guêp. p. 21, pls. 1, 2 (1807).

Merops flavicans, Shaw, Gen. Zool. viii. pt. 1, p. 159 (1811).

Merops apiarius, Steph. in Shaw's Gen. Zool. xiii. pt. 2, p. 173 (1825).

Merops hungariæ, C. L. Brehm, Vög. Deutschl. p. 146 (1831) ; id. Vogelfang, p. 50 (1855).

Merops elegans, C. L. Brehm, Vogelfang, p. 50 (1855).

Guêpier vulgaire, French ; *Abelharuco, Melharuco*, Portuguese ; *Abejaruco*, Spanish ; *Gruccione*, Italian ; *Kirt-elnahal*, Maltese ; *Schegagh*, Arabic ; *el Leeamoon*, Moorish ; *europäischer Bienenfresser*, German ; *Biæder*, Danish ; *Biätare*, Swedish ; *Tschur*, Russian.

Figuræ notabiles.

D'Aubenton, Pl. Enl. 938 ; Werner, Atlas, *Alcyons*, pl. 1 ; Kjærb. Orn. Dan. taf. 13 ; Frisch, Vög. Deutschl. taf. 121, 122 ; Fritsch, Vög. Eur. taf. 14. fig. 1 ; Naumann, Vög. Deutschl. taf. 143 ; Sundevall, Svensk. Fogl. pl. 70. fig. 1 ; Gould, B. of Eur. pl. 59 ; id. B. of G. Brit. ii. pl. 9 ; Dettoni, Ucc. Lomb. pl. 104 ; Dresser, B. of Europe, v. pl. 295.

Hab. Southern Europe, Africa, and South-western Asia.

♂ ad. supra saturatè castaneus, scapularibus et dorso postico toto dilutioribus, fulvescentibus : fronte et superciliis albis viridi-cyaneo lavatis : regione parotica et torque gutturali nigris : tectricibus alarum superioribus ferè castaneis, minimis saturatè viridibus : primariis viridibus, scapis brunneis, pogonio interno versus apicem nigricante : secundariis basin versus castaneis, apice nigricantibus, dorsalibus viridibus : caudâ suprà viridi, subtùs griseâ, scapis rufo-brunneis : gulâ genisque lætè flavis : corpore subtùs cyaneo-viridi, subalaribus fulvescentibus : rostro nigro : pedibus fuscis : iride rubrâ.

Juv. suprà sordidior, dorso postico et uropygio sordidè cyaneo-viridibus, scapularibus cyaneo-griseis vix viridi tinctis : caudâ magis cyaneâ, et rectricibus centralibus nec elongatis : gulâ sordidè flavâ et torque gutturali vix indicato : corpore subtùs sordidè cyaneo-viridi.

Adult male (Seville, 5th May).—Crown, hind neck, and upper part of the back deep chestnut, becoming paler on the back, the colour being deepest in tone on the crown ; a broad frontal band, extending over the eye, white at the base of the bill and otherwise pale bluish green ; back pale chestnut, much lighter on the rump ; upper tail-coverts very pale green ; primary-quills glossy blue-green, tipped with black ; secondaries chestnut, broadly terminated with black, the inner ones washed with bluish green, the elongated innermost secondaries being entirely bluish green ; longer and central wing-coverts chestnut, the smaller ones being bluish green ; scapulars creamy straw-coloured ; tail green with a bluish tinge, the two central rectrices elongated and attenuated ; a line passing below the eye and inclosing the ear-coverts and another across the lower throat deep black ; throat and cheeks rich yellow ; rest of the underparts glossy greenish cobalt, becoming paler on the abdomen and under tail-coverts ; under wing-coverts buff ; under surface of the tail blackish grey ; bill black ; feet pale reddish brown ; iris carmine-red. Total length about 10 inches, culmen 1·6, wing 5·9, tail 5·0, tarsus 0·45.

Adult female (Casa Vieja, 6th May).—Closely resembles the male, but is, if any thing, a trifle duller in colour and a trifle less in size. In a series of specimens, however, these differences do not hold good.

Young (S. Africa, 21st October).—Differs from the adult in having the upper parts duller; the chestnut extends only to the fore part of the back, the rest of the back and rump being dull bluish green; scapulars bluish grey, with a faint greenish tinge; tail more blue in tinge than in the adult and nearly even, the central feathers not elongated; throat of a paler and duller yellow than in the adult, the black band across the lower throat scarcely indicated; rest of the underparts paler and duller than in the adult bird.

Obs. The variations in the series of specimens I have examined are not great, consisting chiefly in the intensity of shade of colour, those from Africa being, as a rule, more richly coloured than examples from Europe. In both males and females the length of the central rectrices varies considerably, and, so far as I can judge, the older birds have these feathers longest. In some specimens the black band below the eye is bordered below with pale turquoise-blue : this is to a slight extent noticeable in a female from Casa Vieja, and much more so in a female from Abyssinia, whereas in an example from the Caucasus there is a very clearly defined broad pale blue line below the black band, and the frontal band is entirely pale blue. The females vary also much in intensity of colour, and in a large series it is impossible to separate the sexes. Mr. Seebohm (Brit. B. vol. ii. p. 324), in his usual anxiety to try and make every one believe that he alone of all naturalists is infallible, says that I "fail to point out the most important characters which distinguish the male from the female." Had he taken the trouble to examine a series instead of blindly adopting Naumann's views (Vög. Deutschl. v. p. 466) respecting the difference in the sexes, he would have seen that in a series the sexes cannot with certainty be distinguished by any external character. I have now before me five carefully sexed females and six males, and find that some of the males have the central rectrices shorter than the females, and in one female from Shiraz they are as long as in any male I have examined. As regards the chestnut on the back, there is a trace of green on several examples of both sexes, and in one female from Abyssinia and another from Spain there is no trace whatever of any green in the chestnut; and, indeed, both these females are as richly coloured as any male in the series. Thus in a series the characters cited by Naumann and Mr. Seebohm, viz. that the female "has the two central tail-feathers not so long, the plumage is not so brilliant, and the chestnut on the back is in many places suffused with green," will not hold good.

I may add that, in both sexes, I find there is a variation in the colour of the tail, it being much greener in some birds than in others, and some examples vary considerably from others in the intensity of shade of colour of the underparts.

In measurements the variation is, on the whole, not much, but African specimens are somewhat smaller than European-killed examples. The variation in size of those I have measured is— culmen 1·5-1·6, wing 5·6-6·1, tail 4·4-5·0, tarsus 0·4.

This, the common European Bee-eater, is generally distributed throughout Southern Europe, and is equally numerous in Northern Africa, ranging even into South Africa, as far south as the Cape Colony, and in Asia it is found as far east as the North-west Provinces of India. As a straggler it has been met with as far north in Europe as Lapland, and has been on many occasions recorded from various parts of the British Isles. According to Professor Newton it has been obtained at least thirty times in Great Britain and four times in Ireland. The first recorded example was exhibited to the Linnean Society by Sir J. E. Smith, and was shot at Mattishall, in Norfolk, in June 1793, and was (*fide* Latham, Syn. Suppl. ii. p. 149) shot out of a flight of about twenty, some survivors of which were observed at the same spot in the following October (Trans. Linn. Soc. iii. p. 333). The specimen was figured by Lewin (Br. B. pl. 43), whose plate is dated Nov. 7, 1793, and having been given by Smith to Lord Derby, is now with the rest of his collection at Liverpool, as its curator, Mr. T. J. Moore, believes. Professor Newton (in Yarr. Br. B. ed. 4, p. 436) recapitulates the records of its occurrences in Great Britain as follows:—"Taking the maritime counties and beginning in the west, Drew states (Hist. Cornwall, i. p. 585) that four were seen and two shot, at Madern near the Land's End, in 1807; while, according to Couch and Bellamy (Nat. Hist. South Devon, p. 202), a flock of twelve, of which eleven were shot, was observed near Helston in the same county in May 1828. In Devon, Dr. Moore in 1837 wrote (Mag. Nat. Hist. ser. 2, i. p. 180) that one was shot at Leigham in April 1818, another at Ivybridge in 1822, and that a third was in Mr. Rowe's collection, while Mr. Nicholl records (Zool. p. 6143) a male killed at Kingsbridge in May 1858. One was shot at Chideock, in Dorset, and preserved in the Bridport Museum, as stated in the first edition of this work. In the Isle of Wight one is said (Zool. p. 4870; Nat. 1855, p. 264) to have been obtained near Freshwater in June 1855. In Sussex, Mr. Knox mentions one shot at Chichester, May 6th, 1829, and Mr. Ellman in 1850 recorded (Zool. p. 2953) one killed at Icklesham, now in Mr. Borrer's collection. As regards Kent, the bird here figured was shot at Kingsgate in May 1827, and another, killed at Lydd, was in 1844 in Dr. Plowley's collection (Zool. p. 623). In Essex, one was killed about midsummer 1854, at Feeting (Zool. p. 4478). Two examples, according to Sheppard and Whitear, were obtained in Suffolk, one at Beccles in the spring of 1825, and the other at Blyburgh in the month of May; while one is supposed to have been seen at Glencham in June 1868 (Zool. p. 1096). In Norfolk, besides the flight in the last century already noticed, Sheppard and Whitear mention one obtained near Yarmouth; Lubbock, in 1845, recorded one from the same neighbourhood more lately, and Mr. Stevenson, in 1866, one killed at Gisleham many years before, in addition to a pair (which there is some reason to suppose had a nest) shot on the river between Norwich and Yarmouth, June 3rd, 1854 (Zool. p. 4367). In Lincolnshire, Mr. Cordeaux notices a specimen without locality or date some years before 1872, and (Zool. 1880, p. 511) a second, shot at Tetney Haven, August 16th, 1880. Further northwards in England there is no satisfactory evidence of this bird's appearance; but Mr. G. R. Gray says he has seen an example obtained in Forfarshire, and one killed in June 1852, at Kinmundy near Peterhead, is recorded (Nat. 1852, p. 204), while Mr. Edward mentions (Zool. p. 6672) a supposed specimen killed about the same time between Hindley and Dufftown. Mr. R. Gray also states (B. of W. of Scotl. p. 513) that a bird believed to be of this species was seen at the close of August 1869 on the river Black Cart in Renfrewshire; and Thompson recorded (Mag. Nat. Hist. ser. 2, ii. p. 18) an example killed October 6th, 1832, near the Mull of Galloway. Mr. Dix, in 1869, mentioned (Zool. s. s. p. 1675) one obtained in Pembrokeshire. Four are said (Zool. s. s. pp. 271 and 561) to have appeared at Stapleton, near Bristol, in April or May 1866, on the 1st or 2nd of which latter month three of them were shot; and forming part, probably, of the same visitation, was a male killed at Bishopstowe in Wilts, May 4th, 1866, as also, according to Mr. Sharpe's information to Capt. Kennedy (B. Berks, &c. pp. 180, 181), one seen for some days in the same year at Dropmore. One shot at Godalming, some years before 1837, rests on the authority of Kidd (Entomol. Mag. iv. p. 270), and Hewitson states (Zool. s. s. p. 2027) that there was an example at Oatlands feeding on yewberries in the late autumn of 1869. In May 1879 a pair was shot near Derby (Zool. 1879, p. 461), one of which is in Mr. Whitaker's collection. As regards Ireland, Vigors many years ago reported (Zool. Journ. i. p. 589) one obtained in the winter of 1820, on the sea-coast near Wexford, in which county Mr. Watters says that another was procured in the summer of 1848; while Dr. J. D. Marshall, in 1829, recorded (Mag. Nat. Hist. ii. p. 395) one killed in Wicklow a few years before, and according to Thompson, Dr. Graves, writing in 1830, mentioned one more obtained in that island, but without giving date or locality."

In Norway, Mr. Collett informs me, the Bee-eater has not been known to have occurred; but it has on several occasions been recorded from Sweden. Professor Sundevall writes that one was shot in June 1816 at Nedraby, a Swedish mile north of Ystad, and was in company with another, which was not secured. One was obtained in Högsäter parish, Dalsland, and presented to the Stockholm Museum by the Rev. O. Fryxell. A small flock of six individuals appeared on the 19th May, 1858, at Täfvelsås, near Wexiö, and remained for three days in a garden. Two of these were shot by the Rev. N. Wieslander, who presented one to the Stockholm, and the other to the Lund Museum. This Bee-eater has even been found within the Arctic circle, as a female was, according to Mr. Meves (Öfv. K. Vet.-Ak. Förh. 1868, p. 264), obtained about a mile south of Muonioniska in Lapland, on the 3rd June, 1865. It has not been observed in Finland, and is of rare occurrence in Northern Russia. Mr. Sabanäeff says that it breeds, though rarely, in the Voronege Government, and it is stated to occur near Moscow. Bogdanoff met with it as far as Samara; Pallas says that it ranges up to the mouth of the Kama river; and according to Eversmann (J. f. O. 1853, p. 291) it ranges northward to Orenberg and the Lower Samara, where the Ik flows into that river. In Southern Russia it is very common and breeds numerously in many localities. According to Mr. Taczanowski it is of accidental and rare occurrence in Poland, and he only knows of one instance of its capture in the Lublin Government, but it has been not unfrequently observed in the Ukraine and Podolia. Borggreve records it from various parts of North Germany, being most common in Silesia, where, according to Gloger, a pair is said to have bred near Ohlau, in 1792. It has been met with in Westphalia and Posen; and Tobias says that it occurs at Hirschberg almost every summer. In Denmark and the Danish Provinces it is a very rare straggler. Mr. Collin states (Skand. Fugl. p. 131) that it has once been met with in Holstein; two were shot on the 5th June 1840, at Gjorslev, in Seeland; and a lady caught two at Klitterne, near Svineklöer. Hornemann states that it has occurred in Fyen; and Mr. Steenberg received one in spirits that had been shot on Anholt in May 1853. It has been obtained on the island of Heligoland; but does not appear to have occurred in Holland or Belgium. It visits Northern France, however; and Messrs. Degland and Gerbe state that a flock of fifteen or twenty individuals established themselves, early in July 1840, at Port Remy, near Abbeville, in a crag already perforated by Sand-Martins, where M. Baillon obtained a sitting female and her eggs. With this exception its occurrences are principally confined to the southern districts, some pairs nesting in Provence every year, although the majority are birds of passage. M. Adrien Lacroix says that it occurs accidentally on passage, from time to time, in the Haute-Garonne. Two were obtained in May 1868 near Portet, about ten kilometres south of Toulouse; and the following year a fine male was shot in April at Saint-Simon, eight kilometres from Toulouse, and was in company with five or six others. Two have, he says, to his knowledge, been obtained in the Hautes-Pyrénées; he received one from Castres in May 1869; and it occurs every year, on passage, more or less numerously, in the Pyrénées orientales.

N

In Portugal the Bee-eater is very common throughout the summer; and in Spain it is exceedingly numerous. I met with it commonly in Catalonia in May; and Colonel Irby writes (Orn. Str. Gibr. p. 65) as follows :—" This bird did not appear to me to be quite so common in Morocco at the end of April as on the Spanish side of the Straits, where during April, May, June, and July it is one of the most conspicuous birds in the country ; at that season Andalucia without Bee-eaters would be like London without Sparrows. Everywhere they are to be seen ; and their single note, *teerrp*, heard continually repeated, magnifies their numbers in imagination. Occasionally they venture into the centre of towns when on passage, hovering round the orange-trees and flowers in some patio or garden. Crossing the Straits for the most part in the early part of the day, flight follows flight for hours in succession. When crossing at Gibraltar they sometimes skim low down to settle for a moment on a bush or a tree, but generally go straight on, often almost out of sight ; but their cry always betrays their presence in the air. My dates of the first arrivals noticed are the 7th of April 1868, 4th of April 1869, 1st April 1870, 29th of March 1871, 26th March 1872, 28th of March 1874. They were observed passing in great numbers from the 10th to the 14th of April in three consecutive years, the greatest quantity arriving on the 10th ; so, in Spanish fashion, I christened that date ' St. Bee-eater's day.' The latest flight I ever saw going north was on the 7th of May. Having remained at Gibraltar once only during July and August, I had but that opportunity of watching the return migration, which appeared during the last week in July and also on the 10th and 12th of August, the last being noticed on the 29th of that month, all, with few exceptions, being heard passing at night. The first arrivals, as is the case with all migrants, are those which remain to breed in the immediate neighbourhood."

This species is recorded by Von Homeyer from the Balearic Islands ; and it is tolerably common all along the coasts of the Mediterranean, but rarer further inland. In Savoy it is only of rare and irregular appearance, being principally met with along the valleys of the Rhône and the Isère ; but in Italy, Sicily, and Sardinia it breeds in many parts, arriving late in April or early in May.

Mr. A. B. Brooke says (Ibis, 1873, p. 236) that in Sardinia, "from about the 17th of April, large flocks began to appear, flying very high in a northerly direction. The first arrivals seemed all to pass on further north ; and it was quite a week later before they began to settle in the south of the island. On their migration they keep up their soft musical note, which can be heard a long distance off. Large numbers breed in the island."

Mr. C. A. Wright states (Ibis, 1864, p. 73), in Malta " it arrives in April and May in large flocks ; and its peculiar gurgling note may be heard at a long distance. Towards evening they settle to roost on the carob-trees, and nestle so close to one another that I have known as many as twenty or thirty to be brought down at one shot. Three were seen in 1861 as late as the 7th June. In Gozo they have been observed to lay their eggs in the sand. They reappear in autumn." In Southern Germany it occurs, as a rule, only as a somewhat rare visitant. Dr. Fritsch speaks of it as being an uncommon bird in Bohemia ; but it is said to have bred there, for Voboril writes that it nested in a vineyard near one of the cemeteries of Prague, and according to Fierlinger it bred on the Pardubic estate some years ago. Specimens have, according to Lokaj, been obtained at Rumburg, according to Palliardi near Prague in 1842, and according to Hromádko in 1847 at the foot of the Kuneticer mountain and near the Forest-house in Raab. Gloger states also that a pair nested near Ohlau in Silesia ; according to Von Homeyer (Naumannia, 1851, p. 65) a nest was found in June 1834 in Württemberg ; and Jäckel states (Naumannia, 1856, p. 152) that one was found some years previously near Würzburg in Bavaria. The late Mr. E. Seidensacher informed me that a specimen of this Bee-eater was obtained near Reichenegg on the 21st May 1864, but that near Marburg it more frequently occurs and has even been seen in small flocks. In Transylvania it is, Messrs. Danford and Harvie Brown state (Ibis, 1875, p. 300), "Local, but, where occurring, found in considerable numbers. Herr Klir saw many during former visits at Bogát, on the Maros, where they were breeding in the river-banks. They seem, however, to be of a wandering disposition; for, although we looked for them at this breeding-place on two occasions, we did not see a single bird, but were always told that they had been there a day or two before. The old nesting-holes which we examined in the low earth-banks of the river were in some instances completely lined with elytra of beetles. Herr Csáto says that in 1850 a great flock appeared at Nagy-oklos, in the Strell valley; and by Bieltz and others Kleinschelken, Birthälen, Nagy-Enyed, Szásváros, &c. are given as localities."

It is common during the breeding-season along the Danube, and breeds in colonies in the banks on many of the southern portions of that river. In Southern Russia and Turkey it is exceedingly numerous during the summer; and according to Messrs. Elwes and Buckley (Ibis, 1870, p. 189) it is common in Turkey, arriving about the same time as the Roller, with which it associates. Colonies of Bee-eaters breed in the earthy cliffs of the Danube, making their holes in the bank like Sand-Martins. Dr. Krüper says that it is a common bird throughout Greece, Macedonia, and Asia Minor, where it arrives early in April, and commences breeding late in May or early in June. He took fresh eggs on the 26th May and the 10th of June in Acarnania, and the 23rd May in Ionia, and incubated eggs on the 8th June on the Isthmus. Eight eggs is the number usually deposited. As soon as the young birds are full-grown they are found in August in flocks on the plains, and leave altogether in September.

Lord Lilford writes (Ibis, 1860, p. 235), "the Bee-eater arrives in Corfu and Epirus in great numbers in April, and breeds in the latter country on the banks of the Kalaito river, near Mursyah, and many other similar localities. In all the holes that we examined, the eggs were laid on the bare sand, without any attempt at a nest. I several times observed three, and once or twice four birds fly from the same hole. These birds leave the country as soon as the young are able to fly. I have never seen them later than the beginning of August. I observed also, in August 1858, on the banks of the Guadalquivir, near San Juan de Alfarache, where there is a large colony of this species, that, although the banks were mined in every direction, and exhibited signs of recent occupation, not a Bee-eater was to be seen."

In Asia Minor it is, as in Greece, very common in all suitable localities; and in Palestine, Canon Tristram writes (Ibis, 1866, p. 83), "though far more numerous in individuals than the Roller, it is less universally distributed, living, however, in large societies in every part of the country. Unlike its smaller congener, *Merops viridis*, it does not frequently perch, but remains for hours on the wing, skimming, Swallow-like, up and down a nullah or wady, or systematically ranging and quartering a barley-plain in pursuit of insects on the wing. Seen athwart the sunbeams as they pass overhead, their colour has the appearance of burnished copper. They feed as well as breed in colonies, preferring low banks to the steeper declivities, and seeming to rely for protection against lizards and other enemies on the structure and turnings of their dwellings rather than on their position. I have taken eggs from a nest in the side of a mere sand-mound on the plain, out of which I started the bird by riding over its hole."

In Africa this Bee-eater is widely distributed, being found from the shores of the Mediterranean to the Cape Colony. Von Heuglin says (Orn. N.O.-Afr. i. p. 197) that it is common

N 2

throughout North-east Africa on passage from the end of March to the beginning of May, and again from August to October, usually in flocks, but not so numerous as *Merops persicus*; and he believes that it breeds in Central Egypt and Arabia Petræa. Captain Shelley says that it arrives in Egypt about the 10th of April, and is then plentifully distributed, but is not quite so abundant as *Merops persicus*. The greater number do not remain to breed, but pass northwards in May, returning in August. In North-west Africa it is also common, and is recorded from Algeria and Morocco as abundant, arriving in the latter country, according to Mr. Tyrwhitt Drake, early in April. On the west coast it has been observed in most localities visited by collectors down to the Cape of Good Hope, where it is said to breed, and it also occurs on the Canaries and Madeira. Vernon Harcourt records it from the latter island; and, according to Dr. Bolle (J. f. O. 1857, p. 324), it frequently visits Fuerteventura in large flocks in winter, and is not unfrequently seen on Canaria. A considerable number are said to have bred on Arguineguin several years in succession. I have examined specimens from many localities on the west coast of Africa; and Mr. Andersson writes (B. of Damara Land, p. 60) that it is " common in Ondonga during the rainy season, when it is also not uncommon in Damara Land proper; but I do not think that it is abundant in Great Namaqua Land.

" These Bee-eaters are observed during their annual migrations in small flocks; but having arrived at their temporary destination they scatter somewhat over the country, though several may still be seen in close proximity. They seem to live chiefly on a species of red wasp, and sometimes seize their food on the wing like Swallows, though they more frequently watch for it from some elevated perch, whence they suddenly pounce upon any prey which may chance to come within their ken, returning invariably to the same spot whether successful or not. When their capture proves a bee or other stinging insect, it is always seized across the body, when the bird, after giving it a sharp squeeze or two between the mandibles of the bill, quickly swallows it. I have seen lizards pursue exactly the same plan when catching hymenopterous insects.

" When on the wing, this Bee-eater utters a pleasant but rather subdued warbling chirp."

Respecting its occurrence in South Africa, Mr. Layard writes (*cf.* Sharpe's ed. of Layard's B. of S. Africa, p. 96) :—" The European Bee-eater, during its period of visitation, extends all over Cape Colony, and we have generally noticed its arrival about August in company with the Quail. All the instances which have been recorded with the actual dates of capture, show that it is only during the months when it is absent from Europe that it visits South Africa, and hence the fact of its breeding during its stay is of great interest. Victorin procured it in the Karroo in January; Dr. Exton shot a specimen at Kanye, and on the 24th of October, 1873, Mr. T. E. Buckley shot two specimens near the river Meathley in Bamangwato. It does not seem to have occurred to Mr. Ayres in Natal, but Mr. Andersson states that it is 'very common in Ondonga during the rainy season, when it is also not uncommon in Damara Land proper;' he considers it to be less abundant in Great Namaqua Land. Señor Anchieta has met with the species on the river Cunene, and also at Caconda in Benguela; while the British Museum contains an example from the Congo, without, however, any indication of the collector.

" It hawks after flies, uttering its cheerful, chirruping cry, and alighting on the summit of the highest bush in the neighbourhood. It breeds in the neighbourhood of Nels Poort, Mr. Henry Jackson having found several nests in holes in banks. The eggs are pure white; axis, 12′″; diam. 10′″. We also found it breeding in great abundance at the Berg river in September and October. It does not always select a bank into which to bore the hole destined for its nest, for we found one flat piece of sandy ground perforated with numberless holes, into which the birds were diving and scrambling like so many rats."

In Asia the Common Bee-eater does not range very far to the eastward. It is found in Arabia; and Mr. Wyatt says (Ibis, 1870, p. 12) that he saw one at El Noweyba, by the Gulf of Akabah, on the 6th April, and on the 9th of that month he found it in abundance near there. Messrs. Dickson and Ross record it (P. Z. S. 1839, p. 119) from Erzeroom, where it arrives about the middle of May, leaving again late in September. Severtzoff states that it breeds commonly in Turkestan; and Mr. Blanford writes (E. Pers. ii. p. 122) that it is "a summer migrant to Persia, and during the warm months it abounds throughout the islands. I met with it first in Baluchistan on the 9th of April; but there, as in Sindh, it is, I suspect, only a bird of passage, and its breeding-quarters are further north; but large numbers undoubtedly remain during the summer and breed in the Persian highlands. The same remarks apply to *Merops persicus* and *Coracias garrulus*, none of these species being found in India in the winter, although they traverse Baluchistan, Sindh, and occasionally North-western India in the spring and autumn; so that it is probable that all of them pass the colder months of the year in Arabia or Africa, and their line of migration crosses at right angles that of such species as *Euspiza melanocephala* and *Coturnix communis*, which resort to India in the winter and breed in the Persian highlands."

Capt. Wardlaw Ramsay writes (Ibis, 1880, p. 49):—" I first observed the European Bee-eater on the 5th June, after which it became quite common in the Hariab valley. On the 22nd of the same month I found it very common between Kurrum fort and the Peiwar Kotal, where neither trees nor shrubs are to be seen for miles. The birds were sitting on the ground and darting up at insects occasionally. . . . . Up to the 10th July, when I left the Kurrum valley, these birds were not breeding; nor, indeed, did I see any place at all suitable for the purpose.

" Surgeon-Major Aitchison, of the Indian Medical Department, the botanist to the Kurrum Expedition, informed me that in a village near the base of the Safed-Koh the villagers said that sometimes in the month of June, when the Bee-eaters arrive, they come down in great numbers to rob the bees from the hives, and that the people had to keep continually on the watch to drive them off. The natives also say that the Bee-eaters do not remain long; so that it is possible that they may go elsewhere to breed."

According to Dr. Jerdon it was observed by Dr. Leith Adams in great numbers in the valley of Cashmere, extending into the plains of the Punjab, and is very abundant at Peshawar. He further remarks (Ibis, 1872, p. 3) that he did not find it so generally spread in Cashmere as he expected, but he saw one immense flock on the Wullur lake in the month of August, evidently about to migrate. According to Pallas it is found in Asiatic Russia only as far east as the Irtish river.

The Bee-eater is certainly one of the most brilliantly coloured and conspicuous of our European birds; and in the rich sunshine of the south there are few more beautiful sights than a flock of these birds hawking after insects. I first saw it alive in Southern Spain, but was too early to find it breeding, and have never been able to take its eggs. Those I saw had just arrived, and were hawking about in pursuit of insects, sometimes resting on the telegraph-wires which passed close to where they were. In their mode of flight they reminded me a good deal of the Swallow, and were catching insects on the wing like that bird. They feed on insects of various kinds, such as bees, wasps, grasshoppers, locusts, and beetles of various kinds, which are chiefly captured on the wing, but are also picked off trees, bushes, or plants. To the beekeeper it is an intolerable nuisance; for one or two of these birds will sometimes watch the entrance to a hive, and almost decimate the bees as they pass and repass.

The present species breeds in colonies in holes in a bank or cliff usually overhanging a stream, but sometimes away from water; and five or eight pure white, glossy, roundish eggs are deposited on the soil in the chamber at the end of the hole. Mr. Osbert Salvin, who met with the Bee-eater breeding numerously in North Africa, writes (Ibis, 1859, p. 303) as follows:— " The first time I observed this species was towards the end of April, at Kef Laks, where a flock, apparently just arrived, passed over my head. It is plentiful about Djendeli, and breeds, boring the hole for its nest, in banks of the river Chemora and the ditches that drain the low land near the lake. There the soil is alluvial and soft, and the bird finds little difficulty in making its excavation. During our stay I took several nests, and latterly became an adept at knowing at once which holes were tenanted, and where and when to dig. A little circumspection is necessary at first; for not unfrequently the occupant of the hole is not a Bee-eater, but a toad or snake. The scratchings made by the bird's feet in passing in and out, and the absence of fresh earth beneath the orifice are generally sure indications of the excavation having been completed, and consequently of a strong probability that there are eggs within. The holes pierced by this bird usually consist of a horizontal passage about three or four feet long, the entrance being at various heights from the level ground. This passage, from a circular opening, is gradually enlarged horizontally till it arrives at a chamber about a foot in diameter, and domed over. In this chamber the eggs are frequently deposited. Should, however, none be found, it is necessary to feel all round the chamber; and in many instances another passage of about a foot long will be found communicating with a second chamber in all respects similar to the first, in which, if it exists, the eggs are placed. The bird makes no nest; but the floor of the chamber is strewn with the legs and wing-cases of Coleoptera in such abundance that a handful may be taken up at once. In most instances I caught one of the old birds in the chamber containing the eggs; while the hole was being enlarged it would, every now and then, attempt to escape. The eggs are laid early in June, and are usually six in number. The flight of the Bee-eater is somewhat like that of a Swallow (*Hirundo rustica*), though its movements are much slower; and it is frequently to be seen perched on a bush. Its cry is harsh and monotonous."

Colonel Irby also (*l. c.*) gives some interesting details respecting the nidification and habits of the present species as observed by him in Southern Spain, where it is very numerous. " Commencing their labours of excavation," he writes, " almost immediately they arrive, the earliest eggs that I know of were taken on the 29th of April; but usually they do not lay till about the second week in May, often not so soon. In some places they nest in large colonies; in others there are perhaps two or three holes. When there are no river-banks or barrancos in which to bore holes, they tunnel down into the ground, where the soil is suitable, in a vertical direction, generally on some slightly elevated mound.

" The shafts to these nests are not usually so long as those in banks of rivers, which sometimes reach to a distance of eight or nine feet in all; the end is enlarged into a round sort of chamber, on the bare soil of which the usual four or five shining white eggs are placed; after a little they become discoloured from the castings of the old birds, the nest being, as it were, lined with the wings and undigested parts of bees and wasps. Vast numbers of eggs and young must be annually destroyed by snakes and lizards: the latter are often seen sunning themselves at the entrance of a hole among a colony of Bee-eaters; and frequently have I avenged the birds by treating the yellow reptile to a charge of shot. The bills of the Bee-eaters, after boring out their habitations, are sometimes worn away to less than half their usual length; but as newly arrived birds never have these stumpy bills, it is evident that they grow again to their original length. It has often been a source of wonder to me how they have the strength to make these long tunnels; the amount of exertion must be enormous; but when one considers the holes of the Sand-Martin, it is not so surprising after all.

"During my stay at Gibraltar, Bee-eaters decreased very much in the neighbourhood, being continually shot, on account of their bright plumage, to put in ladies' hats. Owing to this sad fashion, I saw no less than seven hundred skins, all shot at Tangier in the spring of 1874, which were consigned by Olcese to some dealer in London. However, the enormous injury these birds do to the peasants who keep bees fully merits any amount of punishment; but at the same time they destroy quantities of wasps. After being fired at once or twice they become very wary and shy at the breeding-places; and the best way to shoot them is to hide near the *colmenares* or groups of *corchos* or cork bee-hives, which in Spain are placed in rows, sometimes to the number of seventy or eighty together; and it is no unusual thing to see as many Bee-eaters wheeling round and swooping down, even seizing the bees at the very entrance of their hives.

"Their early departure in August is to be accounted for by the simple fact that bees cease to work when there are no flowers; and by that time all vegetation is scorched up."

As will be seen from the above-quoted notes, the present species deposits its eggs in holes tunnelled in banks by the bird itself, and its eggs, four or five in number, are pure white, very glossy in texture of shell, roundish oval in shape, and in general character much resembling the eggs of the common Kingfisher. Specimens in my collection average in size about $1\frac{1}{40}$ by $\frac{34}{40}$ inch.

The specimens figured are those above described, and are in my own collection.

In the preparation of the above article I have examined the following specimens:—

E Mus. H. E. Dresser.

a, b, ♂ . Barcelona, Spain, 8th May, 1866 (*H. E. D.*). *c, ♂* . Seville, Spain, 5th May, 1868 (*H. Saunders*). *d, ♂* . Gibraltar, 17th April, 1874. *e, ♀* . Casa Vieja, Spain, 6th May, 1874 (*Col. Irby*). *f.* Crimea (*Whitely*). *g.* Caucasus, 19th June, 1871 (*Schmidt*). *h, ♂* . Volga, May 1865 (*Möschler*). *i, k, ♂ ♀* . Sarepta (*Stader*). *l, ♂* . Egypt (*S. Stafford Allen*). *m, ♀* . Saconda, Abyssinia, 21st April, 1868 (*Jesse*). *n.* Damara Land (*Andersson*). *o, ♀* . Shiraz, Persia (*W. T. Blanford*). *p, q.* Cashmere, 1868 (*Jesse*). *r.* Volga (*Dr. Stader*). *s, ♀* . Egypt, 9th April, 1870 (*Shelley*). *t,* juv. S. Africa, 21st October, 1875 (*T. E. Buckley*).

E Mus. Tweeddale.

a, ♀ . Malaga, Spain, 17th May, 1874. *b, ♂* . Black Sea, 9th May, 1866 (*Robson*). *c, ♂* . Carmel, 18th April, 1864. *d, ♂* . Saconda, Abyssinia, 21st April, 1868 (*Jesse*). *e, f, g.* March into Cashmere, 1865 (*S. Pinwill*). *h, ♀* . Cashmere, 9th May, 1876 (*J. Biddulph*). *i, ♂* . Byan Kheyl, 5th June, 1879. *k, ♂* . Byan Kheyl, 16th June, 1879. *l, ♂* . Srinuggur, 21st July, 1876 (*Biddulph*).

E Mus. G. E. Shelley.

a. Southern France (*Shaw Kennedy*). *b, ♂* . Egypt, 3rd April, 1868 (*G. E. Shelley*). *c, ♀* . Nubia, 9th April, 1870 (*G. E. S.*). *d,* juv. Dar-es-Salaam (*Kirk*). *e, ♀* . Caconda, Angola, November 1877 (*Anchieta*). *f.* Cape of Good Hope, 1874 (*Butler*). *g.* Durban, 1874 (*Gordge*). *h, ♂* . Malmsbury, Natal (*I. Van Reenan*). *i, ♂* . Birato, 24th October, 1873 (*T. E. Buckley*). *k, l, ♂* . Bamangwato, 21st October, 1873 (*T. E. Buckley*). *m, n, o, p.* Makalala country (*Bradshaw*).

GREY AND RED BEE EATER

MEROPS MALIMBICUS

MEROPS MALIMBICUS.

GREY-AND-RED BEE-EATER.

Merops bicolor, Daudin, Ann. Mus. Hist. Nat. ii. p. 440, pl. 62. fig. 1 (1803, nec Bodd.) ; Vieill. Tabl. Encycl. et
  Méth. p. 390 (1820) ; Gray, Gen. of B. i. p. 86 (1846) ; Bp. Consp. Gen. Av. i. p. 162 (1850) ; Von Müller,
  J. f. O. 1855, p. 10 ; Schlegel, Mus. Pays-Bas, *Merops*, p. 7 (1863) ; Hartlaub, Orn. Westafr. p. 41 (1857) ;
  Reichenow, J. f. O. 1875, p. 19 ; id. J. f. O. 1877, p. 21 ; Bocage, Orn. d'Angola, p. 89 (1881).
Merops malimbicus, Shaw, Nat. Misc. xvii. pl. 701 (1806) ; id. Gen. Zool. viii. pt. 1, p. 174 (1811) ; Steph. in
  Shaw's Gen. Zool. xiii. pt. 2, p. 75 (1825) ; Lesson, Traité d'Orn. p. 237 (1831) ; Smith, S. Afr. Quart. Journ.
  2nd ser. part ii. p. 319 (1834) ; Ussher, Ibis, 1874, p. 48.
Le Guêpier gris-rose, Levaillant, Hist. Nat. Guêp. p. 28, pl. 5 (1807).
Tephrærops bicolor, Reichenbach, Meropinæ, p. 82 (1852, nec Bodd.).
Tephrærops malimbicus (Shaw), Gray, Hand-l. of B. i. p. 100, no. 1215 (1869).

Figuræ notabiles.

Daudin, Ann. Mus. Hist. Nat. ii. pl. 62. fig. 1 ; Shaw, Nat. Misc. xvii. pl. 701 ; Levaillant, Hist. Nat. Guêp. pl. 5 ;
  Reichenbach, pl. ccclii. figs. 3256, 3257.

HAB. West Africa.

Ad. suprà saturatè schistaceo-cinereus, supracaudalibus vix rubro tinctis : remigibus nigris, intimis apicibus emar-
  ginatis : caudâ rufescenti-fuscâ, schistaceo lavatâ : vittâ suboculari cum regione paroticâ nigris, mento et vittâ
  sub vittâ suboculari albis : gutture et corpore subtùs laetè rosacco-coccineis, abdomine imo cum subcaudalibus
  cinereis : alis et caudâ subtùs sordidè fumoso-cinereis : rostro nigro : pedibus fuscis : iride rubrâ.

Juv. adulto similis, sed caudâ magis cinereo tinctâ et corpore subtùs sordidiore, vittâ albâ in gulæ lateribus angustiore : rectricibus centralibus vix elongatis : iride rufescenti-fuscâ.

Adult male (Abouri).—Entire upper parts dark slaty grey, slightly tinged with red on the
upper tail-coverts ; quills black, the inner ones from the sixth notched at the tip ; tail reddish
brown with a grey tinge ; lores and a stripe passing below the eye and extending over the ear-
coverts black ; chin and a broad stripe bordering the black stripe pure white ; underparts rich
rose-red ; lower abdomen and under tail-coverts ashy grey ; under surface of wings and tail smoky
grey ; bill black ; feet brown ; iris red. Total length about 9 inches, culmen 1·8, wing 5·55,
tail 5·3 (central rectrices extending 1·8 beyond the lateral ones), tarsus 0·5.

Young (West Africa).—Tail much more strongly tinged with grey than in the adult ; underparts duller in colour and the white on the chin and sides of the throat less extensive ; central
tail-feathers but slightly elongated ; iris reddish brown.

THE Grey-and-Red Bee-eater inhabits the western portion of Africa only, and, though not an uncommon bird in collections, but little is known respecting it: all the meagre details that can be gleaned relate merely to its occurrence. First described by Daudin under the name of *Merops bicolor* (a name which properly belongs to an Asiatic species), it has, though so different in appearance, been confused with that species by many naturalists; the first author who gave it an appropriate name was Shaw, who also gave a very good illustration of it in his ‘Naturalist's Miscellany’ (*l. c.*). It appears to range from the Gold Coast down to Angola. Ussher writes (Ibis, 1874, p. 48) that he received two examples from Aubinn, but he does not consider it a common bird in Fantee. Dr. Reichenow (J. f. O. 1875, p. 19) says that he saw several large flocks on the Gold Coast, at the foot of the mountains of Aguapim; they hunted during the day on a steppe, settling on high trees, and in the evening they returned to the mountains for the night. Reichenow also records it from Loanda, Weiss from Elmina, DuChaillu from the Gaboon, and Perrin obtained it in Angola. In habits this Bee-eater does not differ from its allies, and like them it doubtless nests in holes in banks and deposits white eggs; but I find nothing whatever on record respecting its breeding-habits.

The specimens figured and those described are in my own collection.

In the preparation of the above article I have examined the following specimens :—

E Mus. H. E. Dresser.

a. West Africa, 1863 (*DuChaillu*). *b.* Gaboon (*DuChaillu*). *c, ♂* ad. Abouri, 20th February, 1872 (*T. E. Buckley*).

E Mus. Tweeddale.

a, ad. West Africa.

E Mus. Brit.

a. West Africa (*Stevens*). *b, c.* Accra (*Ussher*). *d.* Cape Coast (*Col. Strachan*). *e.* Gaboon (*Verreaux*).

E Mus. Paris.

a. Angola (*Perrin*), type of *Le Guêpier gris-rose* of Levaillant and of *Merops bicolor* of Daudin and Vieillot.

GREEN THROATED SUN EATER

MEROPS NUBICUS.

GREEN-THROATED BEE-EATER.

Le Guêpier à tête bleue, Montb. Hist. Nat. Ois. vi. p. 506 (1779).

Le Guêpier de Nubie, D'Aubenton, Pl. Enl. no. 649.

Merops nubicus, Gmel. Syst. Nat. i. p. 464 (1788) ; Licht. Verz. Doubl. p. 12 (1823) ; Smith, S. Afr. Quart. Journ. 2nd ser. part ii. p. 319 (1834) ; Gray, Gen. of B. i. p. 86 (1846) ; Strickland, P. Z. S. 1850, p. 216 ; Bp. Consp. Gen. Av. i. p. 161 (1850) ; Reichenb. Meropinæ, p. 79 (1852) ; Licht. Nomencl. Av. p. 66 (1854) ; Müll. J. f. O. 1855, p. 10 ; Schlegel, Mus. Pays-Bas, *Merops*, p. 7 (1863) ; Heuglin, J. f. O. 1864, p. 334 ; Hartmann, J. f. O. 1866, p. 202 ; Bocage, Jorn. Sc. Ac. Lisb. ii. p. 135 (1867) ; Hartlaub, Orn. Westafr. p. 41 (1867) ; Heugl. Orn. N.O.-Afr. i. p. 199 (1869) ; Blanf. Geol. & Zool. Abyss. p. 321 (1870) ; Finsch & Hartl. Vög. Ost-Afr. p. 183 (1870) ; Bocage, J. f. O. 1876, p. 435 ; Fischer & Reichenb. J. f. O. 1878, p. 256 ; Fischer, J. f. O. 1879, p. 282 ; Bocage, Orn. d'Angola, p. 90 (1881) ; Forbes, Ibis, 1883, pp. 519, 521, 522, 523, 524–527 ; Shelley, Ibis, 1883, p. 556.

Merops brasiliensis, Gmel. Syst. Nat. i. p. 462 (1788).

Merops cæruleocephalus, Lath. Ind. Orn. i. p. 274 (1790) ; Vieill. Nouv. Dict. xiv. p. 21 (1817) ; Shaw, Gen. Zool. viii. pt. 1, p. 168 (1821) ; Steph. in Shaw's Gen. Zool. xiii. pt. 2, p. 75 (1825) ; Swainson; B. of W. Afr. ii. p. 87, pl. 9 (1837) ; Rüpp. Syst. Uebers. p. 24 (1845) ; Vierthaler, Naumannia, 1857, p. 111.

Merops superbus, Pennant, Ind. Zool. suppl. p. 33 (1790) ; Shaw, Nat. Misc. pl. 78 ; Shaw, Gen. Zool. viii. pt. 1, p. 161 (1811) ; Steph. in Shaw's Gen. Zool. xiii. pt. 2, p. 73 (1825).

Melittotheres nubicus (Gmel.), Reichenb. Meropinæ, p. 82 (1852) ; Cab. Mus. Hein. ii. p. 141 (1859) ; Gray, Hand-l. of B. i. p. 100, no. 1213 (1869).

Figuræ notabiles.

D'Aubenton, Pl. Enl. no. 649 ; Swainson, B. of W. Afr. ii. pl. 9 ; Shaw, Nat. Misc. pl. 78 ; Reichenbach, Meropinæ, pl. cccli. figs. 3254, 3255.

Hab. Northern portion of Tropical Africa.

Ad. capite et collo saturatè viridi-cæruleis : gulâ infrà nigricanti marginatâ : striâ per oculos ductâ et regione paroticâ nigris : dorso et corpore subtùs rubro-sanguineis, hoc vividiore : alis et caudâ rubris sed sordidioribus, remigibus versus apicem viridi lavatis et nigro apicatis : uropygio, supracaudalibus et subcaudalibus cyaneis : rectricibus centralibus valdè elongatis : rostro nigro : pedibus fusco nigricantibus : iride rubrâ.

Juv. adulto similis, sed sordidior, rectricibus centralibus nec elongatis.

Adult (Panganik, E. Africa).—Head and throat deep bluish green, on the throat the blue becomes blackish along the lower margin ; a band through the eye and ear-coverts black ; back and underparts generally rich deep rose-red, or pale carmine with a tinge of vermilion ; wings dull red, the remiges washed with green towards the tip and tipped with black ; rump and upper and under tail-coverts rich blue ; tail dull red, the central rectrices elongated and attenuated towards

o 2

the tip, the terminal portion blackish in colour; under surface of the wing pale fox-red; bill black; legs blackish brown; iris red. Total length about 12 inches, culmen 1·6, wing 5·85, tail 7·5, tarsus 0·45: central rectrices extending about 3½ inches beyond the lateral ones; first and second wing-quills nearly equal and the longest.

Young.—Resembles the adult, but is rather duller in colour and the central rectrices are not elongated.

Obs. So far as I can ascertain from the specimens I have examined, the sexes do not differ in coloration, and in the entire series the difference in measurements is very trifling, viz. culmen 1·3 to 1·6, wing 5·6 to 6·0, and tail 6·5 to 7·6, the variation in the length of the tail being chiefly owing to the difference in the development of the central feathers. The coloration of the throat differs also not a little; for in some examples the blue is very broadly margined below with black, whereas in others there is scarcely any trace of the black margin, and, as a rule, these latter are the smaller birds, though some of them have the central rectrices very long, and it may therefore possibly be a sexual difference.

The range of the present species extends over the northern portions of Tropical Africa. On the west side of the continent it occurs from Senegal down to Bissao, and on the east side from the coast-lands of Abyssinia down to Zanzibar, being replaced in Southern Africa by its ally *Merops nubicoides.* There are specimens in the Leiden Museum from Senegal, and in the British Museum from the river Gambia. Verreaux records it from Casamanze; Forbes met with it on several occasions on the Niger; and I have examined examples from Bissao, below which I do not find any instance of its occurrence on record. On the east side of the continent it is recorded from Abyssinia by Mr. Blanford, who, however, says (*l. c.*) that he only once saw it, when a large number were collected about one shot close to the hot spring of Atfeh, on the shores of Annesley Bay, where Mr. Jesse met with it once. Von Heuglin writes that it is "an inhabitant of the warmer portions of Abyssinia, ascending to an altitude of 6000–7000 feet. We found it common in Takar, Senaar, Kordofan, and along the White Nile." Von Hartmann says (*l. c.*) that "one sees it in Nubia from 16° N. lat. to Kordofan, Senaar, and on the White Nile to the Equator." Dr. Fischer records it (J. f. O. 1879, p. 283) as observed in large numbers in June, north of Mimbrui, on a pasture covered with mimosa trees, and he also met with it at Ngau on the river-banks. I am uncertain as to how far it ranges down in East Africa; but Messrs. Finsch and Hartlaub say (*l. c.*) that according to Von der Decken it is supposed to occur as far south as Zanzibar.

Writing respecting the habits and nidification of this species in North-east Africa, Hartmann says that it is found throughout the year in large flocks, which increase in size during the nesting-season. He met with it in March 1861, where the Amolmul river passes through the south-eastern part of the Djurland, in a flock of about a thousand individuals, which left the river-bank where they were nesting, circled about in the air for a time, and then returned again to the steep river-banks, which they literally carpeted, so that they were coloured with red, green, and blue, and made a lovely picture. Antinori tried to obtain their eggs; but the numerous nest-holes were in the high banks far above the water, and so deep that a two-metre long negro's lance only reached to about two thirds of their depth. Hartmann, after much trouble, got to two nests by digging from above; one contained three and the other four wax-yellow unspotted eggs, which were placed on a few straws and feathers. They measured 23 by 17 millim. These birds retain the rich coloration of their plumage all the year, but specimens obtained in September and October differed from those procured in May by the length of the central rectrices.

Dr. Vierthaler obtained it in September on a journey to Kamlin, and remarks that it was then in full moult. They have a peculiar habit of perching on the backs of Storks (*Ciconia abdimii*); and to see these birds stalking along each with a red rider on his back is a most peculiar sight.

According to Von Heuglin (*l. c.*) it "is gregarious and is occasionally found in flocks of thousands. It breeds about the commencement of the summer rains, in the negro countries along the Abiad as early as in March and April, and in East Sudan between June and August. One finds the breeding-colonies both alongside the water in the forests, in the open parts of the forest-region, and even on the steppes, though here more scattered and frequently in single pairs. The bird digs very deep, usually straight holes, extending either straight or crookedly in the ground; and the nest-chamber is rather enlarged and contains on a slight bed of dry grass-straws three to five eggs, blunt oval in shape, glossy in texture, and pure white in colour; when the egg is fresh it is pale rosy reddish in tinge, but when blown it is yellowish. They measure 11–12''' by 9–9½'''. After breeding, these birds collect in flocks and migrate northwards to about 16° N. lat., passing over the vast savannahs, where they find an abundance of food in the way of grasshoppers. Early in the morning its loud flute-like gurgling call echoes from the bushes and trees where the birds have been roosting, and soon the entire flock flies off and circles about high in the air, uttering their cries, until the dew has dried, when they proceed to hunt after insects in the high grass and along the streams, and, except during midday, they continue hunting most zealously. They live almost exclusively on Orthoptera, which they usually capture on the wing. Should the prairie catch fire these Bee-eaters collect from far and near and shoot in amongst the smoke and flames to seize the insects that are scared up, taking no notice of the numerous birds of prey which have likewise collected in search of food. We frequently saw these birds on the backs of cattle when grazing, especially on asses, and even on the backs of *Ciconia abdimii*, which is a most eager grasshopper-hunter, and is usually seen in small family parties in the high grass. As they walk along they frighten up the smaller Orthoptera, which are caught by the Bee-eaters, who, after securing their prey, settle again on the animals' backs; and I usually observed that the same Bee-eater always returned to the back of the same Stork on which he had previously been riding. During the heat of the day these Bee-eaters take shelter in bushes and trees, which they frequently literally cover, and a flock thus closely settled makes a lovely picture. Between December and April they migrate by degrees southward into the forest-region. In their habits they are, if any thing, more noisy and restless than their allies; their flight is Swallow-like and rapid—they soar and glide along for some distance, then strike quickly with their wings, spread their tail, and rush like an arrow straight up into the air, descending again as quickly. On the Tana lake, in Abyssinia, I saw numbers in February and March, but in April and May they disappeared altogether."

Dr. Brehm speaks of it as being a migrant in the countries watered by the Nile, passing down to about 15° N. lat. during the rainy season, moulting there, and then travelling southward again.

The specimens figured and described are in my own collection.

In the preparation of the above article I have examined the following specimens :—

E Mus. H. E. Dresser.

a, b, c, d, e, f, g. West Africa.   *h*. Bissao (*Verreaux*).   *i*. Panganik, E. Africa (*Kirk*).   *k*. Abyssinia (*Schaufuss*).

E Mus. Paris.

a, b, c. White Nile (*M. d'Arnaud*).   *d*. Bagamoyo (*Mdlle. S. Esprit*).

E Mus. Brit.

a. River Gambia (*Whitely*).   *b*. West Africa (*MacLeannan*).   *c*. Abyssinia (*Blanford*).   *d*, ♀ . Atfoh, Annesley Bay, 5th February, 1868 (*Blanford*).

E Mus. G. E. Shelley.

a, ♂ . Lamo, 1st October, 1877 (*Fischer*).   *b, c*. Lamo (*Kirk*).   *d, e*. Pangani river (*Kirk*).

CARMINE THROATED BEE EATER
FAMILY MEROPIDÆ

MEROPS NUBICOIDES.

CARMINE-THROATED BEE-EATER.

Merops superbus, Vieill. Nouv. Dict. xiv. p. 23 (1816).

Merops nubicoides, Des Murs, Rev. Zool. 1846, p. 243; Bp. Consp. Gen. Av. i. p. 161 (1850); Von Müll. J. f. O. 1855, p. 10; Kirk, Ibis, 1864, p. 324; G. R. Gray, Hand-l. of B. i. p. 100, no. 1214 (1869); Finsch & Hartlaub, Vög. Ost-Afr. p. 185 (1870); Andersson, B. of Damara Land, p. 62 (1872); Gurney, Ibis, 1873, p. 255; Ayres, Ibis, 1874, p. 102; Sharpe, in Layard's B. of S. Afr. 2nd ed. p. 99 (1875-84); Bocage, Jorn. Sc. Ac. Lisb. xxix. p. 63 (1880); Bocage, Orn. d'Angola, p. 537 (1881); Sharpe, in Oates's Matabele Land, p. 301 (1881); Shelley, Ibis, 1882, p. 242.

Melittotheres natalensis, Reichenb. Meropinæ, p. 82 (1852).

Melittotheres nubicoides (Des M.), Bp. Consp. Volucr. Anisodact. p. 8 (1854).

Merops natalensis, Reichenb. Meropinæ, p. 78 (1852); Cab. Mus. Hein. ii. p. 142 (1859); Schlegel, Mus. Pays-Bas, *Merops*, p. 7 (1863).

Merops nubicus, Layard, B. of S. Afr. p. 60 (1867, nec Gmel.).

Inconjani, Matabele.

Figuræ notabiles.

Reichenbach, Meropinæ, pl. ccceli. figs. 3252, 3253; Sharpe, in Layard's B. of S. Afr. pl. iv. fig. 2.

Had. Southern Africa.

Ad. *M. nubico* persimilis, sed mento et gulâ latè coccineis nec cæruleis: rostro nigro: pedibus plumbeo-griseis: iride rubrâ vel rubro-fuscâ.

Juv. pileo sordidè nigricante viridi-cæruleo notato: corpore suprà cum aliis sordidè rufescentibus: remigibus nigro-fusco apicatis, primariis externis in pogonio externo versus apicem viridi-cæruleo lavatis et secundariis intimis cum scapularibus sordidè cæruleo lavatis: caudâ fusco-rufescente, rectricibus centralibus vix elongatis: vittâ per oculos ductâ nigrâ: mento albido, gulâ cæruleâ: corpore cum alis subtùs rufescenti-cervinis: crisso cum subcaudalibus pallidè et sordidè cæruleis: uropygio et supracaudalibus cæruleis.

Adult male (Zambesi).—Resembles the adult of *Merops nubicus* in coloration in every respect except that the entire chin and throat below the black eye-stripe is rich red, rather brighter than the rest of the underparts, instead of blue. Bill black; legs, feet, and claws neutral tint, and feet covered with whitish scales; iris red (*auctt.*), iris dark hazel (*Oates*). Total length about 13 inches, culmen 1·6, wing 6, tail 7·7, tarsus 0·5: central rectrices extending 3·5 beyond the lateral ones; first wing-quills longest, and about 0·35 inch longer than the second.

Adult female.—Resembles the male.

Juv. (Transvaal).—Crown dull blackish, most of the feathers margined and tipped with blue, upper parts generally warm reddish brown with a slight olive tinge, the wings rather more of a dark reddish colour; all the quills tipped with blackish brown, some of the primaries towards the tip externally washed with dull greenish blue, and the inner secondaries also washed with dull blue; rump and upper tail-coverts tolerably bright blue; tail dull brownish red, the central rectrices but slightly elongated; a broad streak through the eye deep black; chin dull white; throat pale blue, a few carmine feathers just showing through the blue; crissum and under tail-coverts pale dull blue; rest of the underparts and under surface of the wing dull reddish buff; under surface of the tail dull blackish grey.

THIS richly coloured bird is found only in the southern portions of Africa; its distribution there appears to be somewhat local, and, though stated to be numerous in some localities, it is still rare, rather than otherwise, in collections.

On the western side of the continent of Africa it appears to range further north than it does on the eastern side; for according to Professor Barboza du Bocage (*l. c.*) it was obtained by Anchieta at Caconda in Angola. Andersson, who records it from Damara Land, writes (B. of Damara Land, p. 62):—" I have only once observed this species, when a specimen occurred a few days' journey south of the river Okavango; its appearance on the wing was beautiful. I understand from the hunters that at certain seasons this Bee-eater is common on the Okavango, and breeds in the banks of that river." In the Cape Colony it is, according to Mr. Sharpe (*l. c.*), of very rare occurrence. M. Jules Verreaux informed him that he obtained a stray specimen at Genadenhal, near Caledon; and one from Natal, which was formerly in Mr. Sharpe's collection, is now in the British Museum. Mr. Ayres forwarded an example to Mr. Gurney, which was, he says (Ibis, 1874, p. 102), sent from the Pindais river, about 130 miles north of Potchefstroom, by Mr. Button. Captain Shelley (*l. c.*) records its occurrence on the Umvuli river, in South-east Africa, where it was obtained by Mr. J. S. Jameson on the 14th September; and, according to this latter gentleman, it "appears in considerable numbers about this date. I am told they breed in some of the banks of the rivers in Mashoona Land. In the Rustenburg district of the Transvaal they are not uncommon."

Dr. Kirk states (*l. c.*) that it was observed both on the Zambesi and the Shiré, and on the former he found colonies of this Bee-eater tunnelling their nests in the river-banks. Mr. Oates also obtained it in the Matabele Land, on the Daka river. I find nothing on record respecting the habits of this Bee-eater beyond the meagre details above cited; and doubtless it does not therein differ from its near ally *Merops nubicus*, being gregarious, like it frequenting open places, river-banks, &c., and burrowing its nest-hole in banks, depositing several glossy pinkish-white eggs.

It is, comparatively speaking, so rare in collections that until quite lately I failed in obtaining an opportunity of examining a young bird. I am, however, now, thanks to Captain Shelley, enabled to give a description of the immature plumage of this Bee-eater; but unfortunately the specimen in question arrived too late for me to have it figured. It is the more interesting because, contrary to what might have been expected, it has the throat pale blue, though one or two of the scarlet feathers of the adult dress are forcing their way through, thus showing that it really is referable to the present species. As this bird is certainly not rare in Southern Africa, it is to be hoped that ere long collectors may send home a better series of specimens than are at present available, and especially birds in the immature dress.

The specimen figured and described is in my own collection.

In the preparation of the above article I have examined the following specimens:—

E Mus. H. E. Dresser.

a, ♂. Zambesi (*Bradshaw*).

E Mus. Brit.

a. Port Natal (*Verreaux*). *b, ♂*. Gcrnah, Matabele Land (*Oates*). *c*. Daka (*Oates*).

E Mus. Tweeddale.

a. South Africa.

E Mus. Paris.

a, b. South Africa, types of *M. nubicoides*, Des Murs (*Delyorgue*). *c*. Port Natal (*Verreaux*).

E Mus. G. E. Shelley.

a, b. Zambesi (*Bradshaw*). *c, ♂ ; d, ♀*. Elands river, Transvaal, February 1883 (*Ayres*). *e, ♀ juv*. Elands river, February 1883 (*Ayres*).

Genus DICROCERCUS.

Merops, Lichtenstein, Cat. rer. nat. rariss. p. 21 (et auctt.) (1793), nec Linn.
Melittophagus, Boie, Isis, 1828, p. 316 (partim).
Dicrocercus, Cabanis, Mus. Hein. ii. p. 136 (1859–60). Type *D. hirundineus.*

HAB. Africa, from Abyssinia and Senegal down to the Cape Colony.

Alis longis, acutis; remige primâ brevissimâ, tertiâ omnium longissimâ, secundâ vix breviore: secundariis longis, intimis primariis æqualibus: caudâ longâ valdè furcatâ: rostro elongato, gracili, curvato: pedibus brevibus, robustis.

Bill long, slender, curved, acute, pentagonal at the base, then four-sided, compressed; gape-line curved; nostrils roundish, nasal membrane short. Wings moderately long, pointed, the first primary shorter than the secondaries, the third longest, the second and fourth somewhat shorter than the third, and nearly equal in length; secondaries rather long, the elongated inner secondaries as long as the primaries. Tail long, deeply forked. Feet small, tolerably stout, the lower portion of the tibia bare; the tarsus scutellate; toes moderately short, rather slender; claws curved, compressed, acute.—Type *Dicrocercus hirundineus.*

————————

THE genus *Dicrocercus* contains but one species, full particulars respecting which are given in the following article.

DICROCERCUS HIRUNDINEUS.

SWALLOW-TAILED BEE-EATER.

Merops hirundineus, Licht. Cat. rer. nat. rariss. p. 21 (1793) ; Finsch & Hartlaub, Vög. Ost-Afr. p. 194 (1870).

Le Guépier à queue fourchue ou le Guépier taiva, Lev. Hist. Nat. Guêp. p. 35, pl. 8 (1807).

Merops furcatus, Stanley, in Salt's Voy. to Abyssinia, App. iv. p. lvii (1814).

Merops hirundinaceus, Vieill. Nouv. Dict. xiv. p. 21 (1817) ; id. Tabl. Encycl. et Méthod. p. 392 (1820) ; Gray, Gen. of B. i. p. 86 (1846) ; Von Müll. J. f. O. 1855, p. 10 ; Hartlaub, Orn. Westafr. p. 40 (1857) ; Schlegel, Mus. Pays-Bas, *Merops*, p. 11 (1863) ; Heuglin, J. f. Orn. 1864, p. 236 ; Monteiro, P. Z. S. 1865, p. 96 ; Bocage, Jorn. Sc. Ac. Lisb. ii. p. 135 (1867) ; Layard, B. of S. Afr. p. 70 (1867) ; Heugl. Orn. N.O.-Afr. i. p. 210 (1869) ; Layard, Ibis, 1869, p. 72 ; Bocage, J. Sc. Ac. Lisb. xvii. p. 50 (1874) ; id. J. f. Orn. 1876, pp. 407, 435 ; id. Orn. d'Angola, p. 93 (1881).

Melittophagus taiva, Boie, Isis, 1828, p. 316.

Merops taiva, Cuvier, Règne Animal, i. p. 442, footnote (1829).

Merops chrysolaimus, Jardine & Selby, Ill. Orn. ii. pl. 99 (1829).

Merops azuror, Less. Traité d'Orn. p. 239 (1831).

Melittophagus hirundinaceus (Vieill.), Bp. Consp. Gen. Av. i. p. 163 (1850) ; Reichenb. Meropinæ, p. 82 (1852) ; Bp. Consp. Volucr. Anisodact. p. 8 (1854) ; Hartmann, J. f. Orn. 1866, p. 204.

Melittophagus furcatus (Stanley), Licht. Nomencl. Av. p. 66 (1854).

Dicrocercus hirundinaceus (Vieill.), Cab. Mus. Hein. ii. p. 136 (1859–60) ; G. R. Gray, Hand-l. of B. i. p. 100, no. 1221 (1869) ; Andersson, B. of Damara Land, p. 63 (1872) ; Ayres, Ibis, 1878, p. 285 ; Nicholson, P. Z. S. 1878, p. 355 ; Sharpe in Layard's B. of S. Afr. p. 101 (1875–84) ; id. in Oates's Matabele Land, p. 302 (1881).

Merops tormisi, auctt. fide Giebel, Thes. Orn. (1875).

Figuræ notabiles.

Levaillant, Hist. Nat. Guêp. pl. 8 ; Jardine & Selby, Ill. Orn. ii. pl. 99.

Hab. Africa, on the eastern side from Abyssinia, and on the western side from Senegal down to the Cape Colony.

Ad. corpore suprà et subtùs lætè psittacino-viridi, pileo et nuchâ vix aurantiaco tinctis : fronte et superciliis turcino-marino-cæruleis : fasciâ transoculari nigrâ : gulâ flavissimâ, fasciâ jugulari pulchrè ultramarino-cæruleâ : remigibus castaneis, extùs viridi lavatis, primariis vix et secundariis valdè nigro terminatis et vix albido apicatis, secundariis intimis elongatis, dorso concoloribus, sed cæruleo lavatis : remigibus centralibus pulchrè cæruleis, reliquis viridi-cærulcis nigro subterminatis et albido apicatis : caudâ valdè furcatâ : abdomine imo, subcaudalibus et supracaudalibus pulchrè ultramarino-cærulcis : rostro nigro : pedibus nigro-fuscis : iride coccineâ.

Juv. (*fide* Heuglin) coloribus dilutioribus : gulâ sordidè cyanescenti-viridi : torque jugulari cyaneo nullo : abdomine magis cærulescenti-viridi : iride lateritiâ.

Adult male (Zambesi).—General colour above and below deep parrot-green, becoming golden in tinge in some lights on the crown and nape; chin and upper throat rich yellow, a broad line through the eyes, extending far back, deep black; forehead, a broad band across the throat, and the upper and under tail-coverts with a portion of the lower abdomen rich dark blue; quills chestnut externally tinged with greenish, the primaries slightly tipped and the secondaries broadly terminated with black; elongated inner secondaries coloured like the back, but tinged with blue; central rectrices deep sky-blue, remaining tail-feathers dull green washed with blue, subterminally marked with blackish, most being tipped with white; bill black; feet dull blackish brown; iris deep red. Total length about 7½ inches, culmen 1·3, wing 3·6, tail 4·1, tarsus 0·4; tail deeply forked, the outer tail-feathers extending 1·5 beyond the central ones.

Adult female (Kabaconeri, 7th July).—Differs but slightly from the male in having the throat of a paler yellow tinge, the lower abdomen and under tail-coverts of a paler blue and tinged with green, the forehead scarcely tinged with blue, and the upper parts rather duller in tinge.

Young (*fide* Heuglin).—Duller than the adult, throat dull bluish green, the blue band across the throat wanting, the abdomen more of a bluish-green tinge.

— — · · · ·

THIS Bee-eater, so easily distinguishable from any other species by its conspicuously forked tail, inhabits Africa, on the eastern side from Abyssinia and the Upper Nile, and on the western side from Senegal down to the Cape Colony. First recorded by Lichtenstein (Cat. rer. nat. rariss. p. 21) in 1793, it was obtained by Salt near Adowa in 1810, and by Stanley described in Salt's Journey (*l. c.*) under the name of *Merops furcatus*. Nothing further appears to have been recorded respecting it during an interval of fifty years, when Antinori met with it on the White Nile in 1860; and Von Heuglin states (*l. c.*) that he first heard of its presence in the country of the Schilluk Negroes, in 12°–14° N. lat., from the travellers Barthélemy and De Pruyssenaere, and subsequently obtained both young and old birds in Wan, Bongo, and on the Kosanga in April, August, September, and October. Hartmann says that it arrives early in February in the Djurland (where it is known by the name of "Adid") and remains until early in April, when it disappears altogether. When passing they are in full breeding-plumage, but soon lose it, as pairs which he shot on the 19th April in the forests of the Dor were not so brilliantly coloured as those he obtained in May. This species frequents overgrown places, and often hunts for food on the edges of forests of high trees, especially near the negro huts, where insect life is abundant. It is very fond of honey, and it is seldom that its beak is not smeared with it. Antinori never saw flocks of more than from eight to ten individuals. Chapman (*teste* Layard) met with it on the Zambesi, and Mr. Nicholson (P. Z. S. 1878, p. 355) received it through Mr. Buxton from Darra-Salam, opposite Zanzibar.

On the west side of the continent it is recorded from Goree by Prof. Barboza du Bocage; there are specimens in the Bremen Museum from Gambia; and Swainson records it from Senegal, Verreaux from Casamanze, Ferguson from Sierra Leone, Pel from Ashantee, Roux from Grand Bassam, Gujon from St. Thomé, Du Chaillu from the Gaboon, Monteiro from Benguela, and Professor Barboza du Bocage received it from Humbe through Anchieta, who states that he met with it on the Cunene river. Andersson states (B. of Damara Land, p. 63) that it was "the commonest species of Bee-eater in Damara Land, and it is also found in Great Namaqua Land and in the Lake country: it chiefly visits Damara Land during the wet season, but a few may be found throughout the year."

Mr. Ortlepp (*teste* Layard, Ibis, 1869, p. 72) observed this Bee-eater in midwinter hawking over the Orange river. Mr. Ayres, who obtained this bird in the Transvaal, says (Ibis, 1878, p. 285):—"I shot a pair in my garden amongst the fruit-trees; they appeared to sit stationary on a bough, and every now and then to dart upon any insect flying past that took their fancy. Their stomachs were well-filled with bluebottles. These are the first birds of the kind I have seen in this part of the country."

The Swallow-tailed Bee-eater is said to resemble *Merops apiaster* in habits; but Mr. Ortlepp says that it does not fly so high when in search of food.

Von Heuglin, speaking of its habits as observed by him in North-east Africa, says (*l. c.*):—"I found this Bee-eater living during and after the rainy season isolated or in pairs in the forest, less frequently amongst bushes. It perches on the highest dry branches of the lofty trees, from whence it takes flight after insects. In April only I saw small flocks of this species, probably on passage; and in August I met with young and old birds together, but they soon separated. According to Antinori it appears on the Djur in March, and remains till early in April (but I shot it there in the end of April), and it then disappears altogether, being still in breeding-plumage, which it soon loses. It is extremely fond of honey, and its bill is frequently plastered with it. It moults in April. Its note resembles that of its allies."

This Bee-eater, like its congeners, deposits its white eggs in a hole tunnelled in a bank. The only account I find respecting its breeding-habits is that given by Mr. Andersson (B. of Damara Land, p. 63), who says:—"I took a nest of this bird on the Omaruru river, on the 31st October. It was situated in a soft sandy bank, some three feet deep horizontally: the entrance was not above two fingers wide; but the hole was slightly enlarged where the nest was found. The nest, which had no lining, contained three beautifully white eggs."

Levaillant says (*l. c.*) that he met with the present species on the banks of the Orange river, and always in localities near water. They were met with by him in isolated pairs; but when the young were fledged they remained with their parents, and roamed about in small flocks of seven or eight individuals. They deposit six or seven eggs of a bluish-white colour, and incubation lasts eighteen days. When leaving, all the families in a district collect and depart together. The cry of this Bee-eater, he says, may be well expressed by the syllable *wi* repeated five or six times in succession; and its name *taiva*, which in the Namaqua language means gall, has been given on account of its greenish-yellow colour, which closely resembles the colour of that substance.

Levaillant further says that this Bee-eater nests also in clefts of rocks and sometimes in holes in trees, a statement that has not been confirmed by any other observer.

This species was undoubtedly first described by Lichtenstein in 1793, as follows (Cat. rer. nat. rariss. p. 21):—"*Merops hirundineus*, nobis. *Merops philippino*, Linn. spec. 5. similis, cauda forficata, jugulo cœruleo. Cf. Buffon, Pl. Enl. n. 57. Probabiliter huc usque ignotus mas *philippini*."

The specimen figured and described is in my own collection.

In the preparation of the above article I have examined the following specimens : —

E Mus. H. E. Dresser.

a, b, ♀. Damara Land (*Andersson*). c, ♀. Kabaconeri, 7th July, 1866 (*Andersson*). d. Transvaal (*Ayres*). e, ♂. Zambesi (*Bradshaw*). f, g, h. South Africa (*Whitely*).

E Mus. Brit.

a, ♀. Damara Land. b, ♀. Port Natal (*Gould*). c. River Gambia (*Strachan*). d, ♂. Victoria Falls, Zambesi, 12th January, 1876 (*Oates*). e. Zanzibar (*Moir*). f, ♂. Potchefstroom, 2nd June, 1877 (*Ayres*). g. Benguela (*Monteiro*). h, ♂. Objimbinque, 2nd July, 1863 (*Andersson*).

E Mus. Paris.

a. Cape of Good Hope (*Verreaux*). b. Senegal (*Mdlle. Rivone*). c. Banks of White Nile (*M. d'Arnaud*).

E Mus. Tweeddale.

a. Lake Ngami (*Chapman*). b. Objimbinque, 23rd November, 1864 (*Andersson*). c, ♀. Objimbinque, 22nd July, 1866 (*Andersson*). d. M'bidgen, Africa, March 1873. e. Rupique, Africa, January 1873.

E Mus. H. Seebohm.

a, ♂. Potchefstroom, 10th June, 1877 (*Ayres*).

Genus MELITTOPHAGUS.

Apiaster, Brisson, Orn. iv. p. 532 (1760, partim).
Merops, Gmelin, Syst. Nat. i. p. 463 (1788, partim).
Melittophagus, Boie, Isis, 1828, p. 316. Type *M. pusillus.*
Meropiscus, Sundevall, Œfv. K. Vet.-Ak. Förh. 1849, p. 162. Type *M. gularis.*
Nyctiornis, Bp. Consp. Gen. Av. i. p. 164 (1850, partim).
Sphecophobus, Reichenbach, Meropinæ, p. 82 (1852). Type *M. pusillus.*
Coccolarynx, Reichenbach, Meropinæ, p. 83 (1852). Type *M. bullocki.*
Urica, Bp. Consp. Volucr. Anisod. p. 8 (1854). Type *M. quinticolor.*
Spheconax, Cabanis, Mus. Hein. ii. p. 133 (1859). Type *M. bullockoides.*
Melittias, Cabanis, Mus. Hein. ii. p. 134 (1859). Type *M. leschenaulti.*

Hab. Ethiopian and Indo-Malayan Regions.

Alis brevibus; remige primâ brevissimâ, tertiâ longissimâ, secundâ et quartâ vix brevioribus: caudâ inconspicuè emarginatâ: rostro longo, curvato, attenuato: juguli plumis nec elongatis: pedibus brevibus, robustis.

Bill long, rather slender, curved, rather stout at the base and attenuated towards the tip, which is sharp-pointed; nostrils roundish, nasal membrane short. Wings moderate, rather short than otherwise, broad, the first quill short, much shorter than the secondaries, the third longest, the second and fourth rather shorter and equal; secondaries long, the elongated inner secondaries very long, extending slightly beyond the fifth quill. Tail moderately long, nearly even, being slightly emarginate, the outer feathers slightly inclined outwards. Feet small, rather stout, the lower part of the tibia sparsely feathered; tarsus scutellate; toes moderate, rather stout than slender, the anterior parallel and partly united; claws moderately slender, curved, compressed, acute.—Type *Melittophagus pusillus.*

The present genus contains eleven species, nine of which inhabit the Ethiopian Region, and the other two (*M. leschenaulti* and *M. quinticolor*) the Indo-Malayan Region.

Like their allies they are gregarious, but not so much so as the species belonging to the genus *Merops*, and are seldom seen in large flocks. They frequent swampy and marshy localities and the banks of rivers, and feed on insects, which they capture on the wing. Their flight resembles that of their allies, and like them their call-note is harsh and monotonous. They nest in holes in the ground, usually near water, and excavate their own nest-holes. Their eggs are roundish, glossy in texture, and pure white in colour.

MELITTOPHAGUS LAFRESNAYEI.

ABYSSINIAN RUFOUS-WINGED BEE-EATER.

Merops variegatus, Rüpp. Neue Wirbelth. p. 72 (1835, nec Vieill.) ; id. Syst. Uebers. p. 24, no. 100 (1845) ; Heugl. Syst. Uebers. no. 143 (1855).

Merops lafresnayei, Guérin, Rev. Zool. 1843, p. 322; id. in Ferret & Galin. Voy. Abyss. iii. p. 243 (1847) ; Brehm, Habesch, p. 210 (1863); Heugl. J. f. O. 1864, p. 355; id. Orn. N.O.-Afr. i. p. 206 (1869) ; Hartl. & Finsch, Orn. Ost-Afr. p. 192 (1870, footnote) ; Finsch, Trans. Zool. Soc. vii. p. 225 (1870) ; Blanf. Geol. & Zool. Abyss. p. 322 (1870).

Merops lefebvrii, Des Murs, Rev. Zool. 1846, p. 243 ; id. Icon. Orn. pl. 34 (1846) ; Des Murs & Prev. in Lefebv. Abyss. Ois. pp. 83, 164, pl. v. (1850).

Melittophagus lafresnayi (Guér.), Bp. Consp. Gen. Av. i. p. 163 (1850) ; Licht. Nomencl. Av. p. 66 (1854) ; Von Müller, J. f. O. 1855, p. 11.

Sphecophobus lafresnayi (Guér.), Reichenb. Meropinæ, p. 82 (1852) ; Gray, Hand-l. of B. i. p. 100, no. 1220 (1869).

Figuræ notabiles.

Ferret & Galin. Voy. Abyss. Atl. pl. 15 ; Des Murs, Icon. Orn. pl. 34 ; Lefebv. Abyss. Ois. pl. v. ; Reichenbach, Meropinæ, pl. ccccxlvii. fig. 3239.

HAB. The highlands of Abyssinia.

Ad. suprà psittacino-viridis, fronte et superciliis lætè ultramarinis : remigibus primariis viridibus intùs versus basin pallidè cinnamomeis : secundariis cinnamomeo-aurantiacis conspicuè nigro terminatis et in apice ipso fusco-albido marginatis : rectricibus duabus centralibus viridibus, reliquis cinnamomeo-aurantiacis conspicuè nigro terminatis et albido apicalis : striâ transoculari nigrâ : gulâ intensè flavâ : torque pectorali latè cyaneo, infrà castaneo cinnamomeo marginato : abdomine reliquo rufescenti-aurantiaco : rostro nigro : pedibus fusco-cinereis : iride rubrâ.

Juv. (fide Heuglin) pallidior, suprà magis cærulescens : gulâ flavicanti-albidâ, lateraliter purius albâ : torque jugulari cyaneo nullo : pectore et epigastrio e cærulescenti-viridi adumbratis.

Adult male (Senafé, 12th May).—Forehead and a broad superciliary stripe rich turquoise-blue, the crown being also marked with the same colour; rest of the upper parts rich parrot-green; primaries green, the basal portions of the inner webs golden orange; secondaries golden orange, broadly terminated with black, narrowly tipped with brownish grey, and marked with green where the orange and black colours meet; elongated innermost secondaries green marked with turquoise-blue; central rectrices green like the back, the remaining tail-feathers deep golden orange, broadly terminated with black and tipped with dull white; a broad black band passes through and behind the eye; chin and throat rich deep yellow, below which a rich blue band crosses the lower throat; rest of the underparts of a deep tawny golden tinge, becoming almost fox-red just below the blue band, and paling to a golden buff on the under tail-coverts; bill blackish; legs brownish grey; iris red. Total length about 7½ inches, culmen 1·3, wing 3·9, tail 3·6, tarsus 0·45.

Adult female.—Resembles the male.

Young (fide Heuglin).—Paler in colour than the adult, the upper parts more blue; throat yellowish white, on the sides pure white, the blue pectoral band wanting; underparts bluish green.

--- --- --- ---

THE present species appears to be restricted to North-east Africa, where it inhabits the more elevated portions of Abyssinia. According to Mr. Blanford (Geol. & Zool. of Abyssinia, p. 322) it is a common species in Abyssinia in the passes from 3000 feet upwards, and was often seen by him in the highlands, especially after the month of March. In the Anseba valley it was replaced by *Melittophagus pusillus.* Mr. Jesse says that it was common up the pass from Sooroo to Senafé during April and May. Von Heuglin says (*l. c.*) that he found this species in the warmer portions of Abyssinia, never in flocks, but in pairs and families. It probably does not migrate, as he saw it in January, February, and March on the Takazié, at Gondar, and in the Galla country, and in July and August in the Samhar and Bogos countries; Lefebvre met with it in August in Schiré, and Brehm at Mensa in the spring. Rüppell speaks of it as a migrant at Gondar in March. It is nowhere common and appears very local, inhabiting localities at an altitude of from 1500 to 7000 feet, and affects high trees along ravines and forest-streams, where it keeps within a limited range, which it leaves unwillingly. In December these birds were fresh moulted, and fledged young were seen directly after the rainy season. Respecting the habits of this Bee-eater I find nothing on record beyond the meagre details above given; and I have no data respecting its nidification, in which it doubtless closely resembles its allies *Melittophagus sonnini* and *M. pusillus.*

The specimen figured and described is in my own collection.

In the preparation of the above article I have examined the following specimens :—

E Mus. H. E. Dresser.

a, ♂. Senafé, Abyssinia, 12th May, 1868 (*Jesse*).   b. Abyssinia (*Verreaux*).

E Mus. Tweeddale.

a. Sooroo, Abyssinia, 5th April, 1868 (*Jesse*).

E Mus. Brit.

a, b. Anseba river, Abyssinia (*Esler*).   c. Abyssinia, 3000 feet, 17th February, 1868 (*W. T. Blanford*).   d. Baraket, Abyssinia, 7500 feet, 24th May, 1868 (*W. T. Blanford*).   e, ♂. Senafé, Abyssinia, 12th May, 1868 (*Jesse*).

E Mus. Paris.

a, b, c, d. Abyssinia (*Petit & Dillon*).

MELITTOPHAGUS SONNINII.

ANGOLAN BEE-EATER.

Le Guêpier à collier gros-bleu ou le Guêpier Sonnini, Levaill. Hist. Nat. Guêp. p. 33, pl. 7 (1807).

Melittophagus sonnini, Boie, Isis, 1828, p. 316; Bp. Consp. Gen. Av. i. p. 163 (1850).

Merops angolensis, Gray, Gen. of B. i. p. 86 (1846, nec Gm.); Reichenow, J. f. Orn. 1877, p. 21.

Melittophagus cyanipectus, Verreaux, Rev. et Mag. Zool. 1851, p. 269.

Merops cyanipectus (Verr.), Reichenbach, Meropinæ, p. 71 (1852).

Sphecophobus sonnini (Boie), Reichenbach, tom. cit. p. 82 (1852).

Sphecophobus cyanipectus (Verr.), Reichenbach, tom. cit. p. 82 (1852).

Melittophagus angolensis, Cab. Mus. Hein. ii. p. 135 (1859-60, nec Gm.).

Merops sonnini (Boie), Bocage, J. Sc. Ac. Lisb. ii. p. 135 (1867); id. op. cit. iv. p. 332 (1869).

Merops variegatus, Bocage, Orn. d'Angola, p. 91 (1881, nec Vieill.).

Figuræ notabiles.

Levaillant, Hist. Nat. Guêp. pl. 7; Reichenbach, Meropinæ, pl. ccccxlvii. figs. 3237, 3238.

Hab. West Africa.

Ad. M. pusillo similis, sed suprà coloribus saturatioribus: fasciâ pectorali lætè cæruleâ nec nigrâ et gulæ lateribus albis facilè distinguendus: rostro et pedibus nigris, iride rubrâ.

Juv. M. pusillo similis, sed major et gulæ lateribus albidis.

Adult (Omoro, Gaboon).—Resembles *M. pusillus*, but has the upper parts rather darker; the band across the lower throat is deep blue, and the sides of the throat below the black band which passes through and behind the eye are pure white. Legs and bill black; iris red. Total length about 6 inches, culmen 1·25, wing 3·4, tail 2·75, tarsus 0·45; tail very slightly emarginate, the outer rectrices being barely longer than the central ones.

Young (Muni river).—Resembles the young of *M. pusillus*, but may be distinguished by being rather larger, and having the black facial patch bordered below with dull white.

It is by no means an easy task to define the range of the present species, as it has by so many authors been confused with its near ally *Melittophagus pusillus* and to some extent also with *Melittophagus lafresnayei*; but, so far as I can ascertain by an examination of the specimens I have been enabled to collect together, it is found only in Western Africa, from the Gambia down to Loango and Angola. In Captain Shelley's collection there is a single example labelled by the

R

late M. Jules Verreaux as having been obtained in Abyssinia; but this locality may well be erroneous, as M. Verreaux was notoriously careless as regards locality when labelling collections that came into his hands in the way of trade. I may add that I have no other reason to believe that it has ever been met with in Abyssinia. Messrs. Finsch and Hartlaub also say that its occurrence in North-east Africa is very doubtful, and the recorded instances are based on the allied species having been mistaken for it. They further state, however, that it has been procured by Kirk in East Africa—which statement I cannot confirm, as amongst the many examples collected by Kirk that I have examined I have not seen one specimen of *Melittophagus sonninii*; and I think that the statement is based on a confusion between the present species and the form of *Melittophagus pusillus* which has the blue superciliary stripe very fully developed, and which has been treated by Cabanis as specifically distinct, under the name of *Melittophagus cyanostictus*, but which does not appear to me to be worthy of specific rank.

Messrs. Finsch and Hartlaub state that the present species is only known with certainty to occur in West Africa, viz. in the Gaboon district, Moondaft, Muni, Cape Lopez, Malimbe, Loanda, and the Rio Chimba. I have examined specimens from the Gambia, Gaboon, and Congo. Dr. Reichenow says (J. f. O. 1875, p. 18) :—"The range of this species is rather restricted in the southern part of West Africa. I obtained it on the Gaboon, but it has not been observed further north. It is a migrant, but does not collect in such large flocks as *Merops albicollis*, and settles on low bushes in preference to trees." He further states (J. f. O. 1877, p. 21) that it occurs in Loanda; and Professor Barboza du Bocage says (Jorn. Sc. Lisb. ii. p. 135) that he received one obtained by Anchieta at Loango, north of the Congo.

In habits this species is stated to closely resemble *M. pusillus*. It lives in pairs or singly, frequenting the vicinity of water, and like its allies feeds on insects, which it captures on the wing. I do not find any thing on record respecting its breeding-habits; but doubtless it makes its nest in a hole in the ground, and deposits pure white eggs.

The present species is very closely allied to *Melittophagus pusillus*, but is sufficiently well differentiated to be entitled to specific rank; for, so far as I can see, the characters by which it is separable are constant. As it has so often been confused with *M. pusillus*, the following table of differences between it and that species, as well as *M. lafresnayei*, may be of use:—

Melittophagus pusillus.	*Melittophagus sonninii.*	*Melittophagus lafresnayei.*
Upper parts parrot-green, rather bright in tinge.	Upper parts darker parrot-green than in *M. pusillus*, and not so bright in tinge.	Upper parts as in *M. sonninii*, but the forehead and crown richly marked with turquoise-blue, and the inner secondaries washed with the same colour.
Has almost always a narrow blue superciliary stripe, which in some specimens is strongly developed.	Has no blue superciliary line.	Has a broad rich turquoise-blue superciliary line.
The band across the throat black, narrowly bordered above with blue.	The band across the throat rich deep blue, and the sides of the throat below the black patch which passes below the eye pure white.	The band across the throat rich turquoise-blue, but rather narrower than in *M. sonninii*.
Culmen 1·1–1·2 in., wing 3·0–3·35, tail 2·45–2·70, tarsus 0·38–0·4.	Culmen 1·2–1·25 in., wing 3·35–3·45, tail 2·75–2·9, tarsus 0·42–0·48.	Culmen 1·3–1·4 in., wing 3·8–3·9, tail 3·35–3·80, tarsus 0·45–0·48.

The specimens figured are those described and are in my own collection.

In the preparation of the above article I have examined the following specimens :—

E Mus. H. E. Dresser.

a, ♂. Gaboon. *b, ♀.* Omoro, Gaboon, 13th June, 1870 (*Skertchley*). *c,* juv. Muni river, West Africa (*Du-Chaillu*).

E Mus. Brit.

a. River Gambia (*Whitely*). *b.* Gaboon (*Verreaux*). *c, ♂.* Omoro, Gaboon, 13th June, 1870 (*Skertchley*). *d.* Gaboon (*DuChaillu*). *e,* juv. Gaboon (*DuChaillu*). *f.* Congo (*Capt. R. M. Sperling*).

E Mus. G. E. Shelley.

a, ♀. Abyssinia? (*Verreaux*). *b, ♂; c, ♀.* Gaboon, 24th June, 1863 (*J. Thomson*).

E Mus. Tweeddale.

a, ad.; *b,* juv. Gaboon (*DuChaillu*).

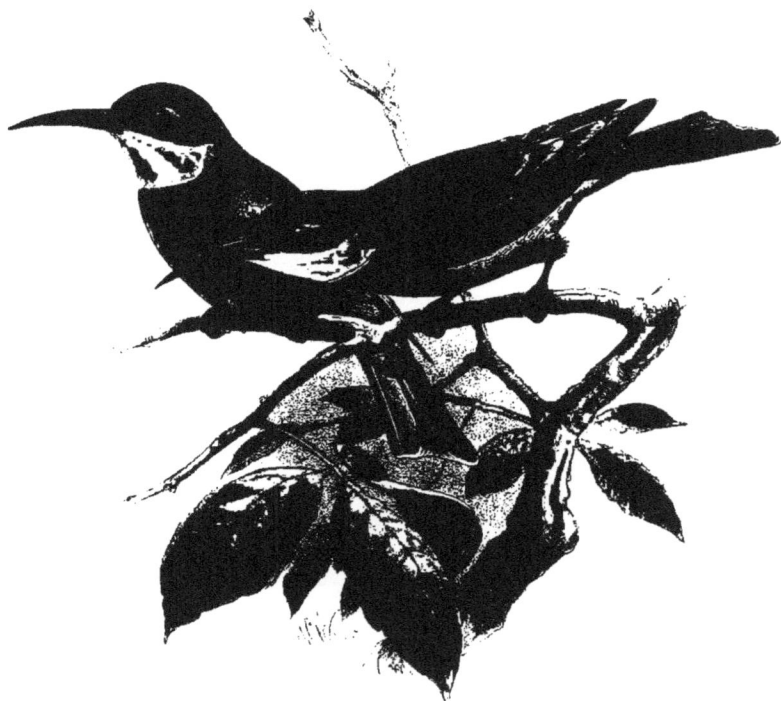

RUFOUS SINGER RED EATER
MELITOGRAPHUS FUSCLUS

MELITTOPHAGUS PUSILLUS.

RUFOUS-WINGED BEE-EATER.

Apiaster angolensis, Briss. Orn. iv. p. 558, pl. xliv. fig. 1 (1760).

Merops pusillus, P. L. S. Müller, Natursystem, Suppl. p. 95 (1776); Blanford, Geol. & Zool. of Abyssinia, p. 322
(1870); Shelley & Buckley, Ibis, 1872, p. 286; Sharpe, P. Z. S. 1872, p. 712; Ussher, Ibis, 1874, p. 48;
Buckley, Ibis, 1874, p. 363; Sharpe in Layard's B. of S. Afr. p. 100 (1875-84); Nicholson, P. Z. S. 1878,
p. 355; Ayres, Ibis, 1879, p. 290; Sharpe in Oates's Matabele Land, p. 301 (1881).

Le petit Guêpier vert et bleu à queue étagée, Montb. Hist. Nat. Ois. vi. p. 503 (1779).

Merops angolensis, Gmel. Syst. Nat. i. p. 463 (1788); Shaw, Gen. Zool. viii. pt. 1, p. 176 (1811); Vieill. Nouv.
Dict. xiv. p. 20 (1816).

Merops erythropterus, Gmel. Syst. Nat. i. p. 464 (1788); Shaw, Gen. Zool. viii. pt. 1, p. 175 (1811); Vieill. Nouv.
Dict. xiv. p. 22 (1816); Licht. Verz. Doubl. p. 13 (1823); Stoph. in Shaw's Gen. Zool. xiii. pt. 2, p. 75
(1825); Rüpp. Syst. Uebers. p. 24 (1845); Strickl. P. Z. S. 1850, p. 216; Hartlaub, Orn. W.-Afr. p. 40
(1857); Schlegel, Mus. Pays-Bas, *Merops*, p. 10 (1863); Heugl. J. f. O. 1864, p. 335; Sclater, P. Z. S. 1864,
p. 111; Monteiro, Ibis, 1865, p. 96; Layard, B. of S. Afr. p. 70 (1867); Monteiro, Ibis, 1869, p. 334;
Heuglin, Orn. N.O.-Afr. i. p. 208 (1869); Sharpe, Ibis, 1869, p. 385; Blanf. Geol. & Zool. of Abyss. p. 322
(1870); Bocage, J. Sc. Ac. Lisb. viii. p. 340 (1870); id. op. cit. xvii. p. 50 (1874); id. J. f. Orn. 1876,
p. 407; id. Orn. d'Angola, p. 92 (1881).

Le Guêpier minute, Levaillant, Hist. Nat. Guêp. p. 53, pl. 17 (1807).

Merops collaris, Vieill. Nouv. Dict. xiv. p. 16 (1816); id. Encycl. Méth. p. 392 (1820); Gray, Gen. of B. i. p. 86
(1846); Von Müller, J. f. O. 1855, p. 10; Hartl. Orn. W.-Afr. p. 40 (1857); Heuglin, J. f. O. 1864, p. 335;
Bocage, J. f. O. 1876, p. 408.

Merops variegatus, Vieill. Nouv. Dict. xiv. p. 25 (1816); id. Encycl. Méth. p. 390 (1820); Hartl. Orn. W.-Afr.
p. 39 (1857); Schlegel, Mus. Pays-Bas, *Merops*, p. 11 (1863); Kirk, Ibis, 1864, p. 324; Finsch & Hartl.
Vög. Ost-Afr. p. 193 (1870); Reichenow, J. f. O. 1875, p. 18.

Melittophagus erythropterus (Gmel.), Boie, Isis, 1828, p. 316; Gray, Gen. of B. i. p. 86 (1846); Bp. Consp. Gen.
Av. i. p. 163 (1850); Licht. Nomencl. Av. p. 66 (1854); Von Müll. J. f. O. 1855, p. 10; Cab. Mus. Hein.
ii. p. 135 (1859); Ayres, Ibis, 1862, p. 27; Hartmann, J. f. O. 1866, p. 205.

Merops minutus, Cuv. Règne Animal, i. p. 442, footnote (1829); Lesson, Traité d'Orn. p. 239 (1831); Finsch, Tr.
Zool. Soc. vii. p. 225 (1870); Finsch & Hartl. Vög. Ost-Afr. p. 191 (1870); Reichenow, J. f. O. 1875, p. 18.

Melittophagus variegatus (Vieill.), Gray, Gen. of Birds, i. p. 86 (1846); Von Müller, J. f. O. 1855, p. 10.

Sphecophobus variegatus (Vieill.), Reichenb. Meropinæ, p. 82 (1852).

Sphecophobus collaris (Vieill.), Reichenb. ut suprà (1852).

Sphecophobus erythropterus (Gmel.), Reichenb. ut suprà (1852).

Merops minultus, A. E. Brehm, J. f. O. 1853, p. 75.

Merops cyanostictus, Cab. in V. d. Decken's Reis. Ost-Afr. iii. p. 34 (1869); Reichenow, J. f. O. 1877, p. 21;
Bocage, J. Sc. Ac. Lisb. xxvii. p. 187 (1880); id. op. cit. xxviii. p. 233 (1880); id. Orn. d'Angola, p. 537(1881).

Melittophagus pusillus (Müll.), Gray, Hand-l. of B. i. p. 101, no. 1222 (1869); Ayres, Ibis, 1871, p. 150;
Andersson, B. of Damara Land, p. 62 (1872).

Melittophagus collaris (Vieill.), Gray, Hand-l. of B. i. p. 101, no. 1223 (1869).
Sphecophobus angolensis (Gm.), Gray, Hand-l. of B. i. p. 100, no. 1219 (1869).
Melittophagus cyanostictus, Cab. J. f. Orn. 1875, p. 340 ; id. op. cit. 1878, p. 335.

Figuræ notabiles.

Briss. Orn. iv. pl. xliv. fig. 1 ; D'Aubenton, Pl. Enl. 318 ; Levaillant, Hist. Nat. Guêp. pl. 17 ; Reichenbach, Meropinæ, pl. cccexlvii. figs. 3240, 3241.

Had. Africa, from Abyssinia and Senegal, down to Natal and the Transvaal.

Ad. suprà psittacino-viridis, remigibus dilutè ferrugineis, primariis extùs viridi marginatis et vix nigricanti apicatis : secundariis valdè nigro terminatis et indistinctè albido apicatis ; secundariis intimis elongatis dorso concoloribus : caudâ ferrugineâ conspicuè nigro terminatâ et albido apicatâ, rectricibus duabus centralibus dorso concoloribus : tæniâ transoculari nigrâ suprà vix cyaneo marginatâ : gulâ flavâ, torque jugulari nigro plerumque cyaneo marginato : pectore lætè fusco-cinnamomeo : abdomine fulvo-aurantiaco vix virescente lavato : rostro et pedibus nigris : iride rubrâ.

Juv. suprà sordidè psittacino-viridis, plumis pallidiore marginatis : gulâ sordidè flavidâ : pectore sordidè viridi : abdomine et subcaudalibus sordidè flavido-cervinis, torque jugulari nullo.

Adult (Fantee).—Upper parts bright parrot-green, not very dark in tinge ; quills bright rufous, externally edged with green, the primaries narrowly, and the short secondaries broadly terminated with black, the latter narrowly tipped with greyish white ; elongated inner secondaries coloured like the back ; tail rufous, broadly terminated with black and narrowly tipped with greyish white, the central rectrices coloured like the back ; chin and throat yellow ; a broad black patch passes from the base of the bill through and behind the eye, and there is a slightly defined blue superciliary stripe ; lower throat crossed by a broad black band, narrowly edged above with blue ; below this band the underparts are rufescent orange, the colour being much darker near the band, and gradually fading towards the vent, which is rufous-buff ; under surface of the wings warm rufous-buff. Bill and legs black ; iris bright red. Total length about 5·5 inches, culmen 1·1, wing 3·2, tail 2·6, tarsus 0·4 ; tail slightly emarginate, the outer feathers extending 0·15 beyond the central ones.

Young (Kakoma).—Upper parts dull parrot-green, the feathers having somewhat paler edges ; throat dull honey-yellow, breast dull green ; abdomen and under tail-coverts dull yellowish buff ; no dark band across the throat.

Although confined to the continent of Africa the range of the present species is widely extended, for it is met with on the east side of the continent from about 16° N. lat., and on the western side from Senegal, down to Natal and the Transvaal, but it does not appear to have occurred within the boundaries of the Cape Colony.

Mr. Blanford found it common in the Anseba valley, Abyssinia, in July. Mr. Jesse procured two specimens at Bojook on the Anseba, but did not observe it elsewhere ; and, according to Von Heuglin, it is one of the commonest of the Bee-eaters in North-east Africa, occurring in Southern Nubia and Takah, and he met with it from the Samhar and Danakil countries into

Abyssinia up to from 7000 to 8000 feet above the sea-level, and along the White and Blue Niles, on the Sabat and Ghazal, westward to Kosanga. It is stated by Cabanis and Heine to occur in Egypt; but Von Heuglin considers this to be an error. Mr. Petherick records it from Kordofan, and Messrs. Fischer and Reichenow from the Galla country. It has been met with in East Africa right down to the Transvaal. Dr. Boehm obtained specimens (which I have now in my collection) at Kakoma and Sagara. Captain Speke shot this bird at Meninga. Sir J. Kirk obtained it at Dar-es-Salaam, opposite Zanzibar; and it has been recorded from Mozambique by Bianconi.

On the western side of the continent of Africa it has been met with, as above stated, from Senegal nearly down to the Cape Colony. Adanson records it from Senegal; Verreaux from Casamanze; there are specimens in the British Museum from Gambia; and it is said to be common on the Gold Coast, whence I have examined many specimens. M. DuChaillu obtained it in Gaboon, and Captain Sperling and others have sent examples from the Congo. According to Dr. Reichenow (J. f. O. 1875, p. 18) it is "common in the river-districts of the Cameroon and Wuri, affecting the maize- and yam-plantations and the open steppes, and frequenting low bushes. I have never seen it high up in the air, and they travel about in pairs or families, and have a more restricted range." It is also, he says, found on the Loango coast. Señor Anchieta procured it on the Cunene river. Professor Barboza du Bocage and Monteiro record it from Angola and Benguela; and the latter, who met with it at Massangano in Angola, remarks that it was "generally seen in the high grass and about flowers, which it searches for insects or honey, and it has a very agreeable chirping song."

In Damara Land, Mr. Andersson writes (B. of Damara Land, p. 62) :—"This exquisite and diminutive species is common on the banks of the rivers Okavango, Teoughe, and Botletlé, as well as on the Lake-watersheds in general, and also about Lake Ngami itself; but I have never observed it so far south as Damara Land proper. It seems to be partial to the immediate neighbourhood of the reedy banks of rivers and of swamps and morasses; and I have never found it at any distance from water." It does not occur within the limits of the Cape Colony, but appears to be common in Natal and the Transvaal. According to Mr. Ayres, who met with it in Natal :— "These Bee-eaters are particularly fond of frequenting reedy marshes and swamps, and are to be found here in certain localities all the year round. They are by no means so plentiful as Savigny's Bee-eater, which is only here in the summer months. It is seldom that more than five or six are to be seen together, and generally not more than two. When feeding, their flight is not so prolonged as that of Savigny's, neither is their note so loud and harsh." He further states that it is abundant about Rustenburg, frequenting sparsely wooded localities, and pretty generally distributed; it is pretty common along the Limpopo, being generally seen in pairs, but sometimes in small companies. Mr. Sharpe (in Layard's B. of S. Afr. p. 100) says that Dr. Exton procured it at Kanye in the Matabili country, and generally throughout Zululand during the winter months, and he wrote to him that "it flies low, and perches on twigs near the ground, from whence it launches after passing insects." Mr. T. E. Buckley writes (Ibis, 1874, p. 363) :—"I saw one or two pairs of this species on the banks of the Limpopo on my way up, and another pair or two on the Samouqui river in the Matabili country. They were plentiful in comparatively open country in the north of the Transvaal on our way down, and were to be seen sitting, singly or in pairs, on a small branch of a bush on the look-out for insects, which they caught on the wing. I once saw a small party of about eight together."

According to Sir J. Kirk, this Bee-eater is widely distributed in the Zambesi country in the vicinity of water.

In its habits this Bee-eater agrees tolerably closely with its allies, but it does not seem to collect in such large flocks as do so many of the other species. Von Heuglin says (*l. c.*) that in North-east Africa it is, "unlike its allies, a resident, and lives in pairs and small families, but is very lively and noisy. It does not rise very high in the air when flying. When the breeding-season is over, it does not wander about the country, and leaves the small district it inhabits most unwillingly. It affects low bushes, thickets overgrown with grass and creepers, cotton-fields, hedges, gardens, and maize-fields, wherever there is water near, and it does not occur in the large dry steppes. On the Gazelle river, I found it inhabiting the swamps, perching like a Kingfisher on the reeds and papyrus-plants, from whence it pursued insects, especially flies. Its note is a guttural flute-like whistle, resembling that of the other Bee-eaters, and at a distance not unlike the call-note of *Limosa melanura*.

"In November and December, I saw several pairs of this Bee-eater on the Asam river, near Adowa, which were flying round the deserted nests of *Hyphantornis larvata*, hanging on acacias, and when I approached they struck down at me. I examined the nests, and in one found two pure white rosy-tinged eggs, measuring 11′″ by 7′″, which were fresh, and which I believe to be those of this species, though König-Warthausen pronounced them to be those of Swifts, or else uncoloured *Hyphantornis* eggs."

I find no authentic account of the breeding-habits of this species on record, but think that in all probability it will be found to breed, like its allies, in holes in the ground, and to deposit pure white glossy eggs. It feeds on insects of various kinds, which it captures chiefly on the wing; and Mr. Monteiro remarks that in the stomachs of specimens that he examined he found remains of small beetles.

The specimens figured are those above described, and are in my own collection.

In the preparation of the above article I have examined the following specimens :—

E Mus. H. E. Dresser.

a. Abyssinia (*Gerrard*). *b,* ♀ ad. Sagara, E. Africa, 21st August, 1880 (*Dr. Böhm*). *c,* juv. Kakoma, E. Africa, 6th February, 1881 (*Dr. Böhm*). *d.* Denkera, December 1871. *e.* Denkera, January 1872 (*Blissett*). *f.* Cape Coast, 1871 (*Ussher*). *g, h, i.* Fantee (*Higgins*). *k, l.* Natal (*Cutter*). *m,* ♂ juv. Transvaal, 29th November, 1873 (*T. E. Buckley*).

E Mus. Brit.

a. Gambia (*Whitely*). *b, c.* Congo (*Sperling*). *d.* Mombas (*R. B. Sharpe*). *e, f.* Mombas. *g.* Tette (*Livingstone*). *h.* Tette (*Oates*). *i.* Denkera (*Blissett*). *k.* Angola (*Monteiro*). *l.* Transvaal (*Ayres*). *m.* Transvaal, 29th November, 1873 (*T. E. Buckley*). *n.* Anseba (*Esler*). *o,* ♂. Bejook, Abyssinia, 14th August, 1868 (*Jesse*).

E Mus. Paris.

a. Senegal (type of *Merops minutus*, Vieillot).

E Mus. Tweeddale.

a. Bejook, Abyssinia, 13th July, 1868 (*Jesse*). *b.* Natal. *c.* Zambesi (*Müller*).

E Mus. G. E. Shelley.

a, b. Fantee (*Ussher*). *c, d.* Accra, 20th February, 1872 (*G. E. S.*). *e, f.* Accra, 5th March, 1872 (*T. E. Buckley*). *g.* Accra, 7th March, 1872 (*T. E. Buckley*). *h.* Durban (*Gordge*). *i,* ♂. Landana, Congo (*Petit*). *k.* Malimbe, Congo (*Petit*). *l,* ♂ juv. Transvaal, 28th November, 1873 (*Kirk*). *m.* Mombas (*Wakefield*). *n,* ♂. Pinetown, 22nd June, 1875 (*Ayres*). *o,* ♀. Rustenburg, June 1877 (*Lucas*). *p.* Dar-es-Salaam (*Kirk*). *q, r.* Dar-es-Salaam (*Buxton*). *s.* Melinda (*Kirk*). *t.* Lamo (*Kirk*). *u, v, w, x, y, z.* Makalala (*Bradshaw*).

EISNE

MELITTOPHAGUS QUINTICOLOR.

CHESTNUT-HEADED BEE-EATER.

? *Merops indicus erythrocephalus*, Brisson, Orn. iv. p. 563, pl. xliv. fig. 3 (1760).

? *Merops erythrocephalus*, Gmel. Syst. Nat. i. p. 463 (1788); Lath. Ind. Orn. p. 274 (1790); Shaw, Gen. Zool. viii.
pt. 1, p. 181 (1811); Gray, Gen. of B. i. p. 86 (1846); Blyth, Cat. Mus. As. Soc. p. 52 (1849); Swinhoe,
P. Z. S. 1871, p. 349; Blyth & Wald. B. Burm. p. 72 (1875).

Le Guêpier quinticolor, Levaillant, Hist. Nat. Guêp. p. 51 (1807).

Merops quinticolor, Vieill. Nouv. Dict. xiv. p. 21 (1816); id. Tabl. Encycl. et Méthod. p. 393 (1820); Bp. Consp.
Gen. Av. i. p. 163 (1850, partim); Kelaart, Prodromus, Cat. p. 119 (1852); Layard, Ann. & Mag. Nat. Hist. xii.
p. 174 (1853); Horsf. & Moore, Cat. B. Mus. E. I. Co. p. 88 (1854); Moore, P. Z. S. 1854, p. 264; Gould,
B. of Asia, pt. viii. (1856); Jerdon, B. of India, i. p. 208 (1862); Beavan, Ibis, 1867, p. 318; Blanf. Ibis,
1870, p. 465; Swinhoe, P. Z. S. 1871, p. 349; Holdsworth, P. Z. S. 1872, p. 423; Jerdon, Ibis, 1872, p. 3;
G. F. L. Marshall, Ibis, 1872, p. 203; Walden, Ibis, 1873, p. 201; Legge, Ibis, 1874, p. 13; Anderson, Zool.
Exp. Yunnan, i. p. 583 (1878); Kelham, Ibis, 1881, p. 377.

Merops urica, Horsf. Trans. Linn. Soc. xiii. p. 172 (1822); Steph. in Shaw's Gen. Zool. xiii. pt. 2, p. 75 (1825);
Less. Man. d'Orn. i. p. 86 (1828); Swains. Classif. of B. ii. p. 333 (1837).

Melittias quinticolor (Vieill.), Cab. Mus. Hein. ii. p. 134 (1859–60); Gray, Hand-l. of B. i. p. 100, no. 1216
(1869).

Merops swinhoei, Hume, Nests & Eggs of Ind. B. p. 182 (1873); id. Stray Feathers, ii. pp. 163, 386 (1874); Ball,
Stray F. ii. p. 386 (1874); Hume, Stray F. iii. p. 50 (1875); Fairbank, Stray F. iv. p. 254 (1876); Armstrong, Stray F. iv. p. 305 (1876); Hume, Stray F. v. p. 18 (1877); Ball, Stray F. vii. p. 203 (1878); Hume,
Str. F. vii. pp. 455, 456 (1878); Hume, Str. F. viii. p. 48 (1879); Tiraut, Bull. Com. Agr. de la Cochin Chine,
sér. 3, vol. i. p. 97 (1879); Vidal, Stray F. ix. p. 49 (1880); Legge, B. of Ceylon, p. 312 (1880).

Merops leschenaulti, Blyth, B. of Burmah, J. As. Soc. Beng. 1875, extra no. p. 72 (nec Vieill.); Hume, Str. Feath.
vi. p. 498 (1878).

Melittophagus leschenaulti, Oates, B. of Brit. Burm. ii. p. 68 (1883, nec Vieill.).

Kurumenne kurulla, Sinhalese, Southern Province; *Pook-kira*, Sinh., N.W. Province (*fide* Legge).

Figuræ notabiles.

Gould, B. of Asia, pl. 13; Reichenbach, Meropinæ, pl. ccccxliii. fig. 3223.

HAB. India, Ceylon, Burmah, Cochin China, China, Malay peninsula.

Ad. fronte, pileo, nuchâ et dorso castaneis: dorso imo cum secundariis intimis psittacino-viridibus: secundariis
intimis cærulco apicatis, remigibus viridibus in pogonio interno rufescentibus et nigro-fusco terminatis:
uropygio et supracaudalibus pallidè cæruleis: rectricibus centralibus in pogonio externo cæruleis et in pogonio
interno viridibus, reliquis viridibus, in pogonio interno fusco marginatis et nigro-fusco terminatis: loris et
maculâ per oculos nigricantibus: mento, gulâ et facie lateribus flavis: torque jugulari castaneo, in parte imâ
nigro et indistinctè flavido marginato: corpore subtùs flavo-viridi, crisso et subcaudalibus cæruleo lavatis:
rostro et pedibus nigris: iride rubrâ.

s

Juv. suprà pallidior, secundariis et tectricibus alarum cærulco marginatis : gulâ albidâ : maculâ oculari indistinctâ et flavido marginatâ : torque jugulari nullo.

Adult male (S. Andamans).—Crown, nape, and interscapulary region bright chestnut-red ; wings and lower back parrot-green, the quills internally margined with rufous and tipped with blackish brown ; the elongated secondaries tipped with blue ; rump and upper tail-coverts light blue, central rectrices bluish on the outer and green on the inner webs, remaining tail-feathers green on the inner web, margined with reddish brown and tipped with blackish brown ; lores and a stripe passing beneath the eye black ; chin and throat rich yellow ; a black band crosses the throat, broadly bordered above with chestnut and below narrowly margined with golden yellow ; underparts bright apple-green, becoming bluish green on the abdomen and under tail-coverts ; bill black ; legs purplish black ; iris scarlet. Total length about 7·5 inches, culmen 1·4, wing 4·4, tail 3·5, tarsus 0·5.

Young (fide Legge).—The chestnut on the upper parts paler in tinge ; wing-coverts and secondaries margined with blue ; throat whitish, the black facial band ill-defined and margined below with yellowish ; no band across the throat, but the lower throat and chest greenish like the rest of the underparts.

THE Chestnut-headed Bee-eater inhabits Southern India, the Andamans, Ceylon, Tenasserim, Burmah, Siam, and Cochin China, and is also stated to have been met with in China, and it ranges down the Malay peninsula as far as Penang. Dr. Jerdon says (B. of India, i. p. 209) :— "This very beautiful Bee-eater, which is the type of the division *Urica*, Bon., is only found in forest-country, and is most abundant in elevated districts. It is found in the Malabar forests and adjoining mountains. I have seen it on the Coonoor Ghât of the Neilgherries up to 5500 feet of elevation ; and it is not uncommon in the Wynaad and other elevated wooded districts. I never saw it on the east coast, nor has it been sent from Central India. Blyth says that it is not found in Lower Bengal, and it is not likely to occur in the North-western Provinces. It extends, however, to Arrakan, Tenasserim, and Malayana." The portions of India whence I find it recorded are the west coast by Bingham, Dehra Doon by Dr. Jerdon, Chota Nagpur by Mr. Ball (who writes, Str. Feath. ii. p. 386, that he met with a pair on the 15th March in the well-wooded hills near the village of Paharbulla in Sirguja), and from N.E. Cachar by Mr. Inglis, who states that it is common there during April and May, but disappears about the end of the latter month.

Mr. Hume remarks that it was not observed in the Nicobars, but it is abundant on the Andaman Islands ; and Mr. Davison says ('Stray Feathers,' ii. p. 163) that "this species is very common in the immediate vicinity of Port Blair, but it is also found, though more sparingly, in the Great and Little Cocos, Strait Island, &c. It is a bird that seldom wanders very far from the forest, and although it is occasionally met with in some extensive clearing, yet it chiefly frequents the roads, running through forest or well-wooded gardens. They breed at the Andamans, and I found them commencing to perforate the banks for their nests just before I left the Andamans in the middle of May." It is by no means an uncommon species in Ceylon. Mr. Holdsworth writes (P. Z. S. 1872, p. 423) that "this is a hill-species, and a resident in Ceylon. I have shot it in August at the foot of the hills in the south, and I have frequently seen it on the lower hills in the neighbourhood of Kandy ; but it is nowhere so numerous as either *M. viridis* or *M. philippinus*, and is generally seen singly or in pairs. I have not observed it on the upper hills. Of two Ceylon specimens, with the chestnut border to the black throat-band, one has the tail entirely green, and the other with the central feathers blue." Col. Legge writes (B. of Ceylon, p. 312):— "This handsome Bee-eater is sparingly dispersed over the island, inhabiting some localities in considerable numbers, while in other districts more stragglers are met with. In the south it is common on the Gindurah river, commencing above Baddegama and extending up into the hills of the Hinedun Pattu; it likewise frequents the banks of the Kaluganga, Kelaniganga, and Maha-oya in the Western Province, and is found here and there through Saffragam. To the north of these localities it is located about Kurunegala, on the Deduru-oya, in the Puttalam district, and in isolated spots in the neighbourhood of Dambulla. Mr. Parker has met with it in the Anaradjapura district, and it occurs sparingly throughout the northern forests. I have seen it between Trincomalie and Mullaittivu, but I do not think it is to be found much to the north of the latter place. In the Kandyan Province it is much more common than in most parts of the low country, inhabiting the vale of Dumbara, Deltota, Nilambe, Maturatta, and Uva generally. It does not ascend to the Nuwara-Elliya plateau."

It is said to be common in Assam, Tenasserim, and Burmah. Mr. Armstrong writes (Str. Feath. iv. p. 305) that "it occurred very sparingly in Southern Pegu. During the months of November, December, and January I did not meet with any specimen of this species, but during the latter end of February I saw several pairs near Elephant Point. They were all remarkably shy, and when disturbed flew away quite out of sight." Mr. Blanford records it as being tolerably common in Pegu and Ava; and according to Mr. Hume it was very common along the forest-streams at the foot of Nwalabo in Tenasserim. Mr. Oates speaks of it as being sparingly distributed throughout British Burmah; and Major Bingham writes to me as follows:—" I first met with this Bee-eater on the western coast of India, on the banks of the forest-streams in the Western Ghâts, and not again till in similarly wooded country I found it common near the rivers in Tenasserim. Though generally to be observed in the vicinity of water, I have more than once come across it in dense forests wherever a break in the jungle afforded clear space for its little flights. I have no pleasanter reminiscences of my wanderings in those eastern forests than those connected with the abundance of this bird, on the oft-repeated marches that I had to make between the large village of Kaukarit on the Houndraw river and the frontier town of Meeawuddy on the Thoungyeen, the river which is the boundary between British Tenasserim and the Shan States. From Kaukarit a winding jungle-road leads along the bed of the Kaukarit stream, straight up to near its source by the Tounjah pass across the Donât mountain-range. Crossing the little stream over thirty times, some of the most beautiful little vistas of forest-scenery open one after the other at each turn of the road. The forest on both sides of the path is dense evergreen—trees crowding on trees on the low steep hillsides, canes and creepers growing in wild luxuriance matting the whole together, so as to render any attempt to stray off the beaten track a work of difficulty. Amidst all this the rivulet winding in and out creates the breaks in the forest, where, on the branches of the trees overhanging its banks, this lovely little Bee-eater can be observed in scores, sitting motionless but watchful, or swooping with cheery whistle at the butterflies which assemble in myriads at the crossings of the stream, making great patches of gorgeous colour on the wet sands. It is wonderfully interesting to watch them dart with sudden sweep at some unwary butterfly passing by, seize it with a loud snap of the bill, and return with an easy, graceful sailing to their perch, on the way deftly shearing off the wings, which flutter unheeded to the ground.

" These birds often descend to the ground, and I have frequently come on one dusting itself like any old sparrow on the roadside. I have also several times seen them clinging on tit-like to the bark of trees, peering into and picking out little insects from the crevices."

To the eastward this Bee-eater ranges into Cochin China. Dr. J. Anderson met with it on the second defile of the Irawaddy ; and M. Tiraut says (Bull. Com. Agric. de la Cochin Chine, sér. 3, vol. i. p. 97) that it is "most common at Trà-vinh, where it occurs with *M. viridis*, though inhabiting different localities ; I never saw it at Thudaû-môt nor at Saigon." There is, I may add, a specimen in the Tweeddale collection from Canton, China.

In the Malay peninsula it is met with down to Penang, whence it is recorded by Mr. Hume (Str. Feath. viii. p. 48) ; and Lieut. Kelham says (*l. c.*) that he shot both the present species and *Merops sumatranus* at Kwala Kangsar, Perak, in February 1877, where they were flying about a sand-bank near the river. I give above some notes from the pen of Major Bingham respecting the habits of this Bee-eater ; and Col. Legge, writing respecting its habits as observed by him in Ceylon, says (B. of Ceylon, p. 313) :—" The banks of rivers which flow through forest or the borders of jungle-begirt tanks are the favourite localities of this bird in the low country. In the Central Province I have seen it principally in the vicinity of rivers in the deep valleys leading to the Mahawelliganga, on roads leading through jungle, and in spots studded with high trees on the sides of steep ravines. It is usually in pairs, and is very arboreal in its habits, sitting on the topmost or most outstretching branches of high trees overhanging water, and darting thence on its prey, much after the manner of a Flycatcher. It takes short flights, and often returns to the same perch again. It is a very pretty object, with its bright green plumage and glistening rufous head, as it darts from the fine old trees lining the forest-rivers down to the edge of the sparkling stream, and glides over the sandy bed, quickly catching up some passing insect. A pair may sometimes be seen seated on a dead twig, touching one another, so very sociable is it in its disposition. It has a soft note, differing from that of either of the foregoing species, which it generally utters from its perch."

Dr. Jerdon says that it "pursues insects from its perch on a lofty tree and generally returns after having captured one. It breeds in holes in banks, generally, but not always, close to water."

Major Bingham writes to me :—" These Bee-eaters dig their holes, so far as I have observed, only in the banks of streams, making a tunnel from four to seven feet deep, inclining downwards at a slight angle. Like that of *Merops viridis* this ends in a rounded chamber, somewhat greater in diameter than the tunnel, which is never lined. The eggs vary from three to five in number, and are a shade larger than those of *M. viridis*, from which they are undistinguishable in colour and shape. In Tenasserim this Bee-eater lays in March, April, and May." Mr. Layard, in his "Notes on the Ornithology of Ceylon " (Ann. & Mag. N. H. 1853, xii. p. 174), says that " the present species affects the hilly forest-region. Here it pursues its insect prey among the lofty tree-tops, seldom descending to the ground, except in the breeding-season, when it frequents steep banks for the purpose of providing a suitable habitation for its young : this is generally effected by scooping a hole in the soil, to the depth of about 18 inches, terminating in a domed chamber, in which the young are hatched on the bare ground. The eggs, two in number, resemble those of the Kingfisher in shape and colour : they are hatched in April."

Col. Legge says (*l. c.*) that he " found the nest of this bird on the banks of the Gindurah in the month of April. The hole was excavated in the soft mould near the top of the bank, went in about 2 feet, with an average diameter of 2 inches, and at the end widened into a cavity of 4 or 5 inches in height and nearly double that in width. There were four young ones lying on the bare ground, which was swarming with live maggots, ants, and flies, brought in for their food by the old birds. The nestlings showed a marked difference in age; two were perhaps not three days old, and the others had the green scapular feathers already sprouting."

I am indebted to Major Bingham for eggs of this Bee-eater, which closely resemble those of *Merops viridis*, but are somewhat larger in size.

The specimens figured are in my own collection.

In the preparation of the above article I have examined the following specimens :—

E Mus. H. E. Dresser.

a, ♂. Tenasserim, 25th October, 1877 (*C. T. Bingham*). b. Houndraw River, Tenasserim, 2nd May, 1879 (*C. T. Bingham*). c. Kota Lama, Perak, 15th February, 1877 (*H. R. Kelham*). d, ♂. S. Andamans, 3rd January, 1873 (*R. G. Wardlaw Ramsay*). e, ♀. S. Andamans, 18th January, 1873 (*R. G. W. R.*).

E Mus. Brit.

a, b. Madras (*Baker*). c. Nepal (*Hodgson*). d. Travancore (*Bourdillon*). e. Ceylon, 22nd February, 1877 (*Legge*). f, g. Tenasserim (*Oates*). h. Lower Pegu (*Oates*). i, ♂. Second defile of the Irawaddy, 5th March, 1875 (*Anderson*). k. Malacca (*Cantor*). l. Penang.

E Mus. Tweeddale.

a. Base of Garo Hills (*Godwin-Austen*). b. Mymensur (*Godwin-Austen*). c. Dehra Doon. d, e. Marachinitly, November 1865 (*S. Chapman*). f, ♂; g, ♀. Burhar, Coonoor Ghât, April 1878 (*Wardlaw Ramsay*). h, ♂. N. Khasia Hills, February 1876 (*A. W. Chennell*). i. Ceylon (*Neville*). k, l. S. Andamans, January 1873 (*Wardlaw Ramsay*). m, n. S. Andamans, 6th February, 1873 (*W. R.*). o, p. Tonghoo, September 1874 (*W. R.*). q, ♀. Rangoon, 30th November, 1873 (*W. R.*). r, ♂. Rangoon, 7th December, 1873 (*W. R.*). s, t. Karen Hills, March 1874 (*W. R.*). u, v. Burmah. w, ♂. Houndraw River, Tenasserim. x. Tenasserim, 6th January, 1877. y, ♀. Tenasserim, 3rd February, 1877. z. Canton, China.

E Mus. H. Seebohm.

a. Bombay Presidency. b. Sikkim, 1877.

E Mus. Paris.

a. Malabar (*Duvaucel*).

JAVAN BEE EATER
MELITTOPHAGUS LESCHENAULTI.

MELITTOPHAGUS LESCHENAULTI.

JAVAN BEE-EATER.

Merops leschenaulti, Vieill. Nouv. Dict. xiv. p. 17 (1816); Gray, Gen. of B. i. p. 86 (1846).

Merops quinticolor, Gray, Gen. of B. i. p. 86 (1846); Bp. Consp. Gen. Av. i. p. 163 (1850, partim); Licht. Nomencl. Av. p. 66 (1854); Schlegel, Mus. Pays-Bas, *Merops*, p. 9 (1863); Nicholson, Ibis, 1881, p. 143.

Melittias quinticolor, Cab. Mus. Hein. ii. p. 134 (1859).

Figura unica.

Levaillant, Hist. Nat. Guêp. pl. 18.

Hab. Java and Sumatra.

Ad. *M. quinticolori* similis, sed capite et dorso vix saturatioribus : caudâ magis cœruleâ, et torque jugulari nigro nec castaneo marginato : rectricibus duabus contralibus vix elongatis.

Juv. adulto similis, sed sordidior : torque jugulari augustiore et saturatè viridi nec nigro.

Adult male (Bantam, Java, 14th June).—Resembles *Melittophagus quinticolor*, but has the chestnut on the head and back rather richer in tinge, the tail is much more blue, and the black band on the throat lacks the rufous margin on the upper part. Total length about 7·5 inches, culmen 1·3, wing 4·0, tail 3·4, tarsus 0·42.

Young (Java).—Resembles the adult, but is duller in general coloration and the band across the lower throat is very narrow and dark green instead of black.

Obs. Besides the differences above cited, I find that in all the Javan specimens I have examined the two central rectrices are slightly elongated, whereas this is not the case in any of the specimens of *M. quinticolor*. In one specimen the central rectrices are fully 0·25 inch longer than the lateral ones.

THE present species is a clearly distinguishable form of *Melittophagus quinticolor*, which species it replaces in Java and Sumatra, its habitat being, so far as can at present be ascertained, restricted to those islands. First recognized as a distinct species by Vieillot and described by him from a specimen sent from Java by M. Laichenot, it has by subsequent authors been very generally united with *Melittophagus quinticolor*, from which species it is, however, very readily separable. To Vieillot the credit is certainly due of having first discriminated between these two species; but though he gave a detailed description of them, showing the distinctive characters, he made a most lamentable mistake in describing the Ceylon species as that occurring in Java, and the Javan one as occurring in Ceylon. There is no doubt that he had before him a specimen from Ceylon and one from Java; but as in all probability they were not labelled, and consequently got transposed, the mistake must have thus occurred. In any case I fully concur with my friend Mr. E. W. Oates, with whom I fully discussed the question when he was writing his 'Birds of British Burmah,' that Vieillot's names for the two species must be retained; but I cannot agree with him that the Indian species should bear the specific name *leschenaulti*, for Vieillot expressly states that he named the Javan bird after the traveller of that name, who obtained it in Java.

Mr. Oates is also in error in surmising that the Javan species was obtained by Lieut. Kelham at Perak, in the Malay peninsula, and was evidently misled by Lieut. Kelham's description. I possess a specimen shot at Perak by that gentleman, which is certainly referable to the Indian and not to the Javan species.

Respecting the habits and nidification of the Javan Bee-eater, I find no data on record; but it doubtless closely resembles its near ally *Melittophagus quinticolor* both in its habits and in its mode of nidification.

The specimens figured are those above described and are in my own collection.

In the preparation of the above article I have examined the following specimens:—

E Mus. H. E. Dresser.

a, ♂. Bantam, Java, 14th June, 1879 (*H. O. Forbes*). b. Java (*Schlegel*). c, juv. Java.

E Mus. Tweeddale.

a, b. Java.

E Mus. G. E. Shelley.

a. Java (*Frank*).

E Mus. H. Seebohm.

a, ♂. Java.

E Mus. Brit.

a, ♂; b, ♀. Java, 1864 (*Diard*).

E Mus. Paris.

a. Sumatra (*Diard*).

BLACK HUIAS

MELITTOPHAGUS GULARIS.

BLACK-BACKED BEE-EATER.

Merops gularis, Shaw, Nat. Misc. ix. pl. 337 (1798); id. Gen. Zool. viii. pt. 1, p. 177 (1811); Vieill. Nouv. Dict.
xiv. p. 16 (1816); Steph. in Shaw's Gen. Zool. xiii. pt. 2, p. 75 (1825); Hartl. Orn. Westafr. p. 42 (1857);
Schlegel, Mus. Pays-Bas, *Merops*, p. 12 (1863); Reichenow, J. f. O. 1875, p. 19.
Melittophagus gularis (Shaw), Gray, Gen. of B. i. p. 86, pl. xxx. (1846); Licht. Nom. Av. p. 66 (1854); Von
Müller, J. f. O. 1855, p. 11.
Meropiscus gularis (Shaw), Sundevall, Öfv. k. Vet.-Ak. Förh. 1840, p. 162; Cab. Mus. Hein. ii. p. 132 (1869);
Sharpe, Ibis, 1869, p. 385; Gray, Hand-l. of B. i. p. 98, no. 1197 (1869); Sharpe, Cat. Afr. B. p. 4 (1871);
Ussher, Ibis, 1874, p. 48; Bouvier & Sharpe, Bull. Soc. Zool. de la France, 1876, p. 305; Bocage, Orn.
d'Angola, p. 94 (1881).
Nyctiornis gularis (Shaw), Bp. Consp. Gen. Av. i. p. 164 (1850).

Figuræ notabiles.

Shaw, Nat. Misc. pl. 337; Gray, Gen. of B. pl. xxx.; Reichenbach, Meropinæ, pl. ccclii. figs. 3258, 3259.

Hab. West Africa.

Ad. niger, fronte, striâ superciliosâ et striâ suboculari turcino-cæruleis et regione paroticâ eodem colore notatâ:
supra- et subcaudalibus cum crisso turcino-cæruleis: corpore subtùs eodem colore striato et guttato: remigibus
primariis duabus externis nigris, reliquis castaneo-rufis, pogonio externo nigro marginatis et terminatis:
secundariis intimis elongatis, nigris, cæruleo marginatis: caudâ nigrâ, rectricibus duabus centralibus turcino
marginatis: mento et gulâ splendidè rubris: pedibus nigro-fuscis: rostro nigro: iride rubrâ.

Juv. saturatè et sordidè niger, dorso vix viridi-cæruleo tincto: uropygio et supracaudalibus sordidè cæruleis: alis
et caudâ sicut in adulto coloratis, sed sordidioribus: mento et gulâ sordidè nigris vix viridi-æneo tinctis et
indistinctè rubro notatis: pectore et corpore subtùs sordidè cæruleo notatis.

Adult male (Fantee).—General plumage deep glossy black; forehead, a superciliary stripe
above, and a slight stripe below the eye, upper and under tail-coverts, lower flanks, and a quantity
of long splashes on the underparts of the body rich turquoise-blue; ear-coverts marked with the
same colour; the primary quills (except the first two, which are black) rich rufous or foxy red,
with the terminal portion and outer web black; most of the short secondaries also foxy red, tipped
with black, the rest black, slightly margined with blue; central rectrices margined with turquoise-
blue; chin and upper throat rich deep vermilion-red; bill black; legs blackish brown; iris red.
Total length about 7 inches, culmen 1·32, wing 3·7, tail 3·0, tarsus 0·42.

Adult female (fide Reichenow).—Resembles the male, but when freshly killed has the black
on the upper portions of the plumage tinged with green.

T

Young (Fantee).—Deep, rather dull black, the back glossed to some extent with greenish blue; rump and upper tail-coverts dull cobalt-blue; wings and tail as in the adult but duller; chin and upper throat glossed with coppery green, a single red feather showing here and there; breast and rest of the underparts marked with dull cobalt-blue.

THE Black-backed Bee-eater inhabits the west coast of Africa only, where it is met with from Sierra Leone down to Angola. Judging from the numbers of specimens which have been sent in collections from the Gold Coast, it appears to be very abundant in that region; and Governor Ussher, who collected there for some time, writes (Ibis, 1874, p. 48):—"This very beautiful little Bee-eater is tolerably common in Fantee and is occasionally seen in company with *M. albicollis*; but whereas the latter species keeps very low, affecting small bushes or the lower branches of trees, the present bird invariably selects the highest vantage-point it can find, a naked branch in preference to a leafy one, from which it makes occasional sallies after its prey. I have never observed more than three or four together, whereas *M. albicollis* is sometimes met with in very large numbers at a time. The vicinity of water appears to be selected by *M. gularis* in preference to any other situation." Dr. Reichenow found it generally distributed on the Gold Coast, and says (J. f. O. 1875, p. 19) that he met with it in all parts of the west coast of Africa visited by him, from the Gold Coast to the Gaboon; and DuChaillu sent many specimens from the latter locality, where it is evidently a common species; below this it appears to become more sparingly distributed. Petit obtained it at Louembo Chissambo in the Congo district; and, so far as I can ascertain, there is but one record of its presence in Angola, viz. that of a specimen in immature dress sent by Mr. A. de Fonseca from Cazengo in Angola to Professor Barboza du Bocage (*cf.* Orn. d'Angola, p. 94).

But little is recorded respecting the habits of this Bee-eater, and I have never succeeded in procuring its eggs or in obtaining any reliable data respecting its breeding-habits; and from what Dr. Reichenow (whose notes I give below) says, it appears probable that it nests in hollow trees. If this should prove to be the case, it will be a most interesting fact, as tending to show that it differs materially in its nesting-habits from any of its allies. Indeed, Dr. Reichenow remarks (*l. c.*) that he found it to "differ considerably from its congeners in its habits." "I never met with it," he says, "on the open steppe, but wherever high bushes and trees grow together or cover large districts, or else in the open places of the large forests, it is to be found, and it also ascends to a considerable height in the mountains. I always found it resident in pairs or singly, and it does not seem to migrate; in fact it has no reason to do so like its allies. I believe that this Bee-eater breeds in holes in trees, for I often saw males in the forests, whose mates were probably busy with the cares of nidification, and there were no suitable places for that purpose but holes of trees anywhere within reach. In fresh-killed specimens the female has the back of a lighter blackish green, whereas in the male this part is almost pure black. In skins this difference soon disappears."

The specimens figured are in adult plumage and are, together with those described, in my own collection.

In the preparation of the above article I have examined the following specimens :—

E Mus. H. E. Dresser.

a, b, c, d, e, ad. Fantee, Gold Coast (*Ussher*). *f,* juv. Fantee, Gold Coast (*Ussher*). *g.* Wassau, Gold Coast (*Swanzy*).

E Mus. Brit.

a. Fantee (*Ussher*). *b, c.* Gold Coast (*Capt. Moloney*). *d.* Ashantee (*G. Lagden*). *e.* Wassau (*Capts. Burton & Cameron*). *f.* Gaboon (*Gould coll.*). *g.* Gaboon (*Walker*).

E Mus. Tweeddale.

a, b. Gaboon (*DuChaillu*).

E Mus. Paris.

a. Gaboon (*Aubry Lecomte*). *b.* Coast of Africa between Cape Palmas and the Calabar coast (*Laurein*).

REDTHROATED BIL BEEATER
MELITTOPHAGUS MUELLERI

MELITTOPHAGUS MUELLERI.

RED-THROATED BLUE BEE-EATER.

Meropiscus mülleri, Cassin, Journ. Ac. Sci. Phil. 1857, p. 37; Cab. Mus. Hein. ii. p. 132 (1859) ; Gray, Hand-l.
of B. i. p. 98, no. 1198 (1869).
Merops mülleri (Cass.), Hartlaub, Orn. Westafr. p. 262.
Meropogon mülleri, Cassin, Journ. Ac. Sci. Phil. ser. 2, iv. p. 322, pl. 49. fig. 2 (1860).
Nyctiornis mülleri (Cass.), Giebel, Thes. Orn. ii. p. 734 (1875).

Figura unica.

Cassin, Journ. Ac. Sci. Phil. ser. 2, iv. pl. 49. fig. 2.

Hab. Gold Coast.

Ad. capite et collo cum corpore subtùs, suprà- et subcaudalibus saturatè cæruleis, vix nigro notatis: striâ per oculos
ductâ et regione paroticâ nigris : gulâ vividè coccineâ : dorso cum alis suprà saturatè ferrugineis, remigibus
saturatioribus : caudâ nigrâ, saturatè cæruleo lavatâ : rostro nigro : pedibus nigro-fuscis : iride rubrâ.

Adult (Fantee).—Head, neck, underparts generally, and the under and upper tail-coverts rich
cobalt-blue, the black of the basal portion of the feathers showing through here and there, more
especially on the head and nape; a streak through the eye and the ear-coverts black; throat-
patch rich deep vermilion-red, verging on carmine-red; back and wings deep foxy rufous, the
quills rather darker; tail black, washed with cobalt-blue; bill black; legs blackish brown; iris
deep red. Total length about 6·5 inches, culmen 1·32, wing 3·18, tail 2·0, tarsus 0·4: first quill
short, nearly an inch shorter than the second; third and fourth longest; second shorter than fifth.

It is a somewhat difficult matter to write the history of a bird about which so little is known as
the present species. First obtained by DuChaillu on the Muni river, in West Africa, it was for
long only known from there, the single example obtained by him being for some time unique;
but subsequently one example was sent from Fantee to Mr. R. Bowdler Sharpe, and is now in
the British Museum; and a third specimen came in a small collection from the Gold Coast to
Capt. G. E. Shelley. Beyond these three I do not know of any specimens in existence, and all
being, so far as we can judge, in adult dress, the immature plumage of this rare Bee-eater is
unknown. It appears to be most nearly allied to *Melittophagus gularis*, and probably closely
resembles that species in its habits; but I find nothing whatever on record respecting its general
habits, note, or mode of nidification.

I have not been able to procure a specimen for my own collection, and the bird above described and figured is in the British Museum.

In the preparation of the above article I have examined the following specimens :—

E Mus. Brit.

a. Fantee (*Sharpe coll.*).

E Mus. G. E. Shelley.

a. Gold Coast.

WHITEFRONTED BEE EATER
MELITTOPHAGUS BULLOCKOIDES

MELITTOPHAGUS BULLOCKOIDES.

WHITE-FRONTED BEE-EATER.

Merops bullockoides, Smith, S. Afr. Quart. Journ. 2nd ser. part ii. p. 320 (1834); id. Ill. Zool. S. Afr., Aves, pl. ix. (1849); Kirk, Ibis, 1864, p. 324; Layard, B. of S. Afr. p. 70 (1867); Bocage, J. Sc. Ac. Lisb. ii. p. 135 (1867); id. op. cit. v. p. 48 (1868); Sharpe, Cat. Afr. B. p. 4 (1871); Buckley, Ibis, 1874, p. 363; Bocage, J. Sc. Ac. Lisb. xvii. p. 35 (1874); id. J. f. O. 1876, p. 407; Sharpe & Bouvier, Bull. Soc. Zool. de la France, 1878, p. 475; Ayres, Ibis, 1879, p. 289; Sharpe in Oates's Matabele Land, App. p. 301 (1881); Bocage, Orn. d'Angola, p. 93 (1881); Sharpe in Layard's B. of S. Afr. p. 99 (1875–84).

Melittophagus bullockoides (Smith), Gray, Gen. of B. i. p. 86 (1846); Bp. Consp. Gen. Av. i. p. 164 (1850); Licht. Nomencl. Av. p. 66 (1854); Von Müller, J. f. O. 1855, p. 11; Ayres, Ibis, 1865, p. 265; Gray, Hand-l. of B. i. p. 101, no. 1225 (1869); Ayres, Ibis, 1871, p. 150.

Merops bullockioides (Smith), Reichenbach, Meropinæ, p. 77 (1852).

Coccolarynx bullockioides (Smith), Reichenb. Meropinæ, p. 83 (1852).

Spheconax albifrons, Cab. Mus. Hein. ii. p. 133 (1859–60).

Merops smithi, Schlegel, Mus. Pays-Bas, *Merops*, p. 9 (1863).

Merops albifrons (Cab.), Finsch & Hartl. Vög. Ost-Afr. p. 188 (1870); Giebel, Thes. Orn. ii. p. 571 (1875).

Figuræ notabiles.

Smith, Ill. Zool. S. Afr., Aves, pl. ix.; Reichenbach, Meropinæ, pl. cccci. figs. 3248, 3249; Sharpe in Layard's B. of S. Afr. pl. iv. fig. 1.

Hab. Southern Africa.

Ad. fronte, mento et striâ infraoculari albis: nuchâ aurantiaco-cervinâ: corpore suprà cum alis et caudâ psittacino-viridibus: primariis intimis inconspicuè nigro apicatis et secundariis (intimis elongatis exceptis) valdè nigro terminatis: loris cum striâ magnâ per oculos ductâ nigris: jugulo sanguineo: corpore subtùs aurantiacocervino: crisso cum suprà- et subcaudalibus lætè azureis: rostro et pedibus nigris: iride coccineâ.

Juv. ubique pallidior et sordidior: jugulo pallidè rubro, et crisso cum subcaudalibus pallidè azureis.

Adult male (Makalaka Country).—Forehead silvery white; crown pale brownish, intermixed with white; nape and hind neck pale orange-buff; upper parts generally, including the upper surface of the wings and tail, rich parrot-green, slightly tinged with bluish green; inner primaries slightly and secondaries (excepting the elongated innermost ones) broadly terminated with deep black; upper and under tail-coverts and crissum deep cobalt-blue; lores and a broad stripe passing through and behind the eye deep black; chin and a broad stripe below the black stripe pure white; throat bright scarlet; rest of the underparts warm golden buff, becoming blue on the lower abdomen; bill and legs blackish; iris red. Total length about 8·5 inches, culmen 1·65, wing 4·6, tail 3·85, tarsus 0·58.

Adult female.—Does not differ in plumage from the male.

Young (Rustenberg).—Differs from the adult in being much paler in coloration, in having the throat pale reddish and not deep red, and the crissum and under tail-coverts pale blue.

THE White-fronted Bee-eater inhabits the southern portion of the continent of Africa, but has been recorded from as far north as the Gaboon on the western side of that continent by Du-Chaillu, who met with it at Cape Lopez. Messrs. Sharpe and Bouvier record its capture at Malimbe. Welwitsch obtained it in Angola, and according to Professor Barboza du Bocage it has been procured by Anchieta on the river Cunene at Huille, in Mossamedes and Humbe, but it does not appear to have ever been observed by Andersson in Damara Land. It was first described by Sir A. Smith, who writes (*l. c.*):—"It was not until the expedition attained the 25° of south latitude that this bird was discovered, though north of that it appeared not uncommon. When observed, it was generally either perched upon the tops of trees, along the immediate banks of rivers, or in the act of making short circuits through the air, apparently in chase of flying insects. As may be inferred from the structure of its wings, it is not a bird which flies for any great length of time without resting; it seeks its food during frequent low and short flights, and after each of these often returns to the perch from whence it proceeded. In respect of its habits, as well as its wings, it closely resembles *Merops erythropterus*, Linn.; but, in regard to both these characters, it differs from the other species of the genus yet observed in South Africa. Upon the modified structure of the wings in this species and *Merops erythropterus* may depend the circumstance of their being permanent inhabitants of the districts where they are found, and where they encounter a cold during the winter much more severe than ever occurs farther to the southward, and from which the *Merops apiaster*, Linn., *Merops savignii*, Levaill., and *Merops chrysolaimus*, Jard., fly towards the end of summer. From observations I have had occasion to make, I think it probable that the migrations, both of birds and quadrupeds, will be found often to depend more upon causes which have hitherto been comparatively overlooked, than upon any absolute deficiency of food in the countries from whence they retire. Connected with this opinion, I may instance the circumstance of a species of Swallow, which inhabits the mountains of the Cape Colony during the summer months, repairing in the winter to the vicinity of houses, left by another species on the approach of the cold season. It there finds food sufficient for its support, till the other species, gifted with more vigorous power of flight and a superior courage, returns and drives it back again to situations which it had for a time abandoned."

Mr. Layard says (B. of S. Afr. p. 70) that he received several specimens through Mr. David Arnot of Colesberg, which were procured in the Orange Free State. Mr. T. E. Buckley (Ibis, 1874, p. 363) found this species breeding on the banks of the Limpopo, perhaps seven or eight pairs in one colony, and he adds that he found it common throughout the north of the Transvaal. Mr. Ayres records it as exceedingly common round Rustenberg in the Transvaal, and he also procured it from the Monocusi river in Natal. Oates met with it on the Crocodile river in Matabele Land; and Livingstone (Miss. Travels, p. 248) found this Bee-eater, and also *Merops apiaster*, breeding in society in holes in the banks of the river Locambye in November and December. It is found on the east coast as far north as the Zambesi, where it was obtained by Sir J. Kirk.

In its habits this Bee-eater is said to closely resemble its allies; it frequents river-banks and the vicinity of water, is usually seen perched on the tops of trees, and like its allies feeds on insects of various kinds, which it captures on the wing. Sir J. Kirk states that it is solitary in its habits and frequents the banks of streams; Mr. Ayres found them in flights on the banks of rivers, generally alighting on the tops of bushes and trees or on any bare exposed twig, and he observes that their notes are harsh and short. "Towards evening near Rustenberg," the latter writes (Ibis, 1879, p. 289), "they go in flights, appearing to congregate and roost at certain known localities, generally on the sides of a gully with perpendicular banks, on the ledges of which they sleep; in such situations they also breed during the summer months, as is evident from the many holes bored in the banks; during the day they generally disperse, and may then be found solitary, or but two or three together."

This Bee-eater breeds in holes in river-banks, and deposits roundish, pure white, glossy eggs, similar to those of its allies.

The adult bird figured and described is in my own collection, and the young bird described is in the British Museum.

In the preparation of the above article I have examined the following specimens:—

E Mus. H. E. Dresser.

a, ♂. Transvaal (*Ayres*). b, c. Transvaal (*Barratt*). d, ♂; e, f. Port Natal (*Cutter*). g. Makalaka Country (*Bradshaw*).

E Mus. Brit.

a, ♂. Transvaal (*Ayres*). b, juv. Near Rustenberg (*Barratt*). c. Natal (*Cutter*). d. Crocodile river (*Oates*).

E Mus. Tweeddale.

a. Zambesi (*Kirk*).

E Mus. Paris.

a. South Africa, 1845 (*Delagorgue*). b. Moszelekatœ Country, South Africa, 1836 (*Verreaux*).

E Mus. G. E. Shelley.

a. Makalaka Country (*Bradshaw*). b, c, d. Rovumak (*J. Thomson*).

BLUE BROWED GREEN BEE EATER
MEROPS MEGLATENSIS

MEROPS MUSCATENSIS.

BLUE-BROWED GREEN BEE-EATER.

Merops muscatensis, Sharpe, Ibis, 1886, p. 15.

Figura nulla.

Hab. Muscat.

Ad. corpore suprà sicut in *M. viridi* colorato, sed striâ frontali et superciliari pallidè viridi-cæruleâ : gulâ et gutture sicut in *M. cyanophryi* coloratis, sed paullò pallidioribus et clarius coloratis : plagâ nigrâ minore : abdomine sicut in *M. viridi*, sed magis cæruleo tincto : rostro nigro : pedibus sordidè plumbeis : iride scarlatinâ.

Adult.—Upper parts as in *Merops viridis*, but with a narrow frontal and superciliary stripe of a pale greenish-blue colour ; throat and breast as in *Merops cyanophrys*, but the blue on the throat is rather clearer and lighter in tinge of colour, is rather less extended, and the black patch is somewhat smaller ; rest of the underparts as in *Merops viridis*, but somewhat more tinged with blue ; beak black ; legs dull dark plumbeous ; iris bright red. Total length about 6·75 inches, culmen 1·12, wing 3·6, tail 3·9, tarsus 0·4 : central rectrices extending 0·7 beyond the lateral ones.

JUST as I had sent to the printers the last batch of MS. for the present work, my friend Mr. R. Bowdler Sharpe wrote asking me to come and see an apparently new Bee-eater which had been sent to the British Museum from Muscat by Col. Miles. I at once took my two specimens of *Merops cyanophrys* for comparison, and found the bird in question to be a fairly good species, exactly intermediate between *Merops viridis* and *Merops cyanophrys*. Mr. Sharpe has named it in the January number of 'The Ibis' for 1886, and a detailed description will be given by him in the April number of that periodical. Compared with Persian examples of *Merops viridis* and my two specimens of *Merops cyanophrys*, the present species agrees with the former in the coloration of the upper parts and abdomen, but it has a clearly defined narrow greenish-blue frontal and superciliary stripe, and the central rectrices are much shorter and blunter. It agrees more closely with *Merops cyanophrys* in the coloration of the throat and breast, but the blue is rather paler and clearer and the black patch is smaller. In size it agrees more closely with *M. viridis* than with

M. cyanophrys; in a word it is just intermediate between the two species, and sufficiently distinct from either to warrant specific distinction.

Nothing is at present known respecting its range, except that it occurs at Muscat, and no notes were sent respecting its habits. The only specimen I have seen is that in the British Museum collection, which is the one I have figured and described.

MELITHREPTUS JULARIS
(YOUNG)

BLUE VENTED BEE EATER
MELITTOPHAGUS BULLOCKI

MELITTOPHAGUS BULLOCKI.

BLUE-VENTED BEE-EATER.

Le Guêpier à gorge rouge ou le Guêpier bulock, Levaillant, Hist. Nat. Guêp. p. 59, pl. 20 (1807).

Merops bulocki, Vieill. Nouv. Dict. xiv. p. 13 (1817); id. Tabl. Encycl. et Méth. p. 391 (1820); Rüpp. Syst. Uebers. p. 24 (1845); Vierthaler, Naumannia, 1852, i. pp. 40, 44, 53.

Merops bullocki (Vieill.), Lesson, Traité d'Orn. p. 237 (1831); Schlegel, Mus. Pays-Bas, *Merops*, p. 8 (1863); Sharpe & Bouvier, Bull. Soc. Zool. France, 1876, p. 304; Forbes, Ibis, 1883, p. 533.

Merops bullockii (Vieill.), Smith, S. Afr. Quart. Journ. 2nd ser. part ii. p. 320 (1834); Reichenbach, Meropinæ, p. 78 (1852); A. E. Brehm, J. f. O. 1853, pp. 455, 456, Extrah. p. 97; Hartlaub, Orn. W.-Afr. p. 41 (1857); Heugl. Orn. N.O.-Afr. i. p. 204 (1869).

Merops cyanogaster, Swains. B. of W. Afr. ii. p. 80, pl. 8 (1837).

Melittophagus bullocki (Vieill.), Gray, Gen. of B. i. p. 86 (1846); Bp. Consp. Gen. Av. i. p. 163 (1850); Lichtenstein, Nomencl. Av. p. 66 (1854); Von Müller, J. f. O. 1855, p. 11; Shelley, Ibis, 1883, p. 556.

Coccolarynx bullocki (Vieill.), Reichenb. Meropinæ, p. 83 (1852); Gray, Hand-l. of B. i. p. 101, no. 1224 (1869).

Spheconax bullocki (Vieill.), Cab. Mus. Hein. ii. p. 134 (1859–60).

Spheconax frenatus (Hartl.), Cab. ut suprà.

Merops frenatus, Hartlaub, J. f. O. 1854, p. 257; Heuglin, J. f. O. 1864, p. 335.

Melittophagus bullockii (Vieill.), Hartmann, J. f. O. 1866, p. 204.

Coccolarynx frenatus (Hartl.), Gray, Hand-l. of B. i. p. 101, no. 1226 (1869).

Figuræ notabiles.

Swainson, B. of W. Afr. ii. pl. 8; Reichenbach, Meropinæ, pl. cccel. figs. 3250, 3251.

HAB. Northern portions of East and West Africa.

Ad. suprà psittacino-viridis, pilei plumis cyaneo angustè apicatis: nuchâ et collo postico rufescenti-cervinis, dorso supremo eodem colore lavato: remigibus primariis vix nigricante apicatis, secundariis (intimis elongatis exceptis) conspicuè nigro terminatis, colore nigro ipso postico, basin remigum versus, lætè cyaneo lavato: rectricibus virescente cinnamomeis, duabus centralibus dorso concoloribus: tæniâ transoculari nigrâ cyaneo marginatâ, et striâ frontali cyanoâ: gulâ scarlatinâ: pectore abdominoque aurantiaco-cervinis: abdomine imo et subcaudalibus saturatè ultramarinis: rostro nigro: pedibus fuscis: iride fusco-rubrâ.

Adult (Senegal).—Crown, upper surface of body, wings, and tail parrot-green, the feathers on the crown slightly tipped with pale cobalt-blue; nape and hind neck warm rufescent-buff, the fore part of the back tinged with the same colour; primary-quills slightly tipped with dull blackish, the secondaries, all but the elongated inner ones, broadly terminated with deep black, and crossed at the base of the black by a narrow blue line; tail rufescent orange tinged with green, the two central rectrices being, however, green like the back; lores and a patch extending through the eye over the ear-coverts deep black, slightly bordered with turquoise-blue, there being also an indistinct turquoise-blue line over the forehead; chin and throat rich glossy carmine-red;

x

rest of the underparts rich golden buff, except the lower abdomen which, with the under tail-coverts, is rich deep cobalt-blue ; bill black ; legs dark brownish ; iris deep brownish red. Total length about 8 inches, culmen 1·3, wing 3·9, tail 3·68, tarsus 0·45.

THE range of this Bee-eater is restricted to Eastern and North-western Africa, being replaced in Southern Africa by *Melittophagus bullockoides*. Von Heuglin met with it at Sennar ; Vierthaler obtained it at Abu Schok on the 24th December, and found a breeding colony on the banks of the Nile, but did not obtain its eggs ; Antinori records it from the Blue Nile ; and, as stated below, Von Heuglin met with it on the Wau river, in East Central Africa, in April.

On the western side of the African continent it is recorded from Senegal and the river Gambia ; Petit obtained it at Malimbe, and Mr. Forbes met with it near Shonga on the Niger.

Antinori considers this Bee-eater to be a migrant on the Blue Nile and not to be found there from December to February ; but Von Heuglin and Brehm say that it breeds there from December to March. According to Von Heuglin (*l. c.*) this bird " has a very limited range in North Africa. Rüppell, and probably also Ferret and Galinier, found it in the north-eastern lowlands of Abyssinia ; and we met with it on the Mareb at Hamedo, on the western slope of the Abyssinian highlands, especially near the rain-beds which have their outfall in the Deuder, Rahad, and Atbara, on the upper Blue Nile in Fazogl, and, lastly, on the Wau river in Central Africa, in April. Except in the breeding-season only one pair or one family are met with together ; and they are to be seen in the forenoon and afternoon in the groves, or near water, sitting watching for insects on the dead boughs of bushes or low trees, or else on high grasses or plants. They are less noisy and restless than their allies." In October 1859, when Dr. Hartmann, together with the late Dr. Pency, ascended the Blue Nile, the latter shot a pair from an *Acacia nilotica* at one discharge : these were the last they saw that season, for they were migrating further towards the equator. In March 1860, Dr. Hartmann, when settled at Rosêres, saw several small flocks which remained there until late in April : early in the morning and at sunset they were busy hunting up insects from the ground ; but during the heat of the day, when some pairs had retired to the shade of the palms, most of them left for the higher places above Rosêres, where they hunted after insects amongst the gum-trees which grew plentifully there. He found in the stomachs of these birds honey-eating insects. The plumage of this species, especially the red neck-spot, is much richer in colour in March and April than in October, and its feathers are then much fresher.

On the western side of the African continent this Bee-eater is by no means uncommon, as many specimens have been sent from Senegal and the river Gambia. Petit obtained it at Malimbe ; and Mr. Forbes, who met with it near Shonga on the Niger on the 17th December, says that it was observed in some numbers, settling in high trees on the banks and flying off in true Bee-eater style. So far as I can ascertain, there is no authentic instance of *Melittophagus bullocki* having occurred in South Africa.

Like its allies this Bee-eater nests in holes which it excavates in the banks of rivers. Vierthaler found it breeding near Abu Schok on the Nile in December, but did not obtain its eggs ; and Dr. A. E. Brehm writes (J. f. O. 1853, Extrah. p. 97) as follows :—" Of the Bee-eaters I only found breeding colonies of *Merops bullockii*. A party of forty to sixty pairs select a smooth firm bank on the shores of the Blue Nile to excavate their nest-holes, ut such a place is always selected in the densest forest, no doubt because the bird finds most food there. The holes, which have a diameter of about 1½ inch, and are 3 to 5 feet deep, terminate in an oven-shaped enlargement which forms the nest, and which measures 2½ to 3 inches high, 4 to 6 inches broad, and 6 to 8 inches long. The entire colony, consisting of forty to sixty holes, covers a space of 25 to 30 square feet, the holes being about 4 to 6 inches apart. What an unerring eye these birds must have for each to distinguish its own hole amongst so many, for when flying in they only remain a second before the holes! I found such breeding-colonies on the 24th December 1850, and on the 13th, 24th, and 27th January, and 11th February 1851, but was never fortunate enough to find eggs or even nest-materials."

After a careful examination of specimens, as stated below, I have failed to find any specific difference between specimens from East and West Africa. Dr. G. Hartlaub, in proposing to separate the North-east African bird from the form occurring in West Africa, writes (J. f. O. 1854, p. 257) as follows:—" The supposition that the West-African *Merops bullockii* occurs also in the central and north-eastern provinces of Africa, seems to require verification. Numerous examples of so-called *M. bullockii* from Sennar, compared by us, differ constantly from the West-African bird in having a turquoise-blue stripe, which, passing from below the chin, borders the black patch on the side of the head, and on the forehead is clearly defined to above the eye. This characteristic is present in both sexes, and is wanting entirely in the true *M. bullockii* of West Africa, which, as correctly stated by Swainson, has only a slight tinge of blue before the nostrils. Besides, the dark blue edging to the black wing-spot is richer and clearer in the East-African bird (which I propose to call *M. frenatus*) than in *M. bullockii*. The former is a larger bird, three examples now before me are fully 8½ Parisian inches long. Levaillant, Latham, Vieillot, and Swainson describe the West-African bird. Reichenbach (Meropinæ, p. 78) was the first and only author who figured and described our new species, which he considered to be *M. bullockii*, and incorrectly considered that the characteristics had escaped the notice of the authors named. With regard to the geographical distribution of these two closely allied species, *Merops bullockii* appears to be confined to the west coast of Senegambia ; at least proof of its occurrence in Guinea appears to be wanting. In the British Museum there is a specimen stated to have come from South Africa (*cf.* Gray, List, etc. part ii. sect. 1, p. 72) ; we doubt if the locality is correct, as in South Africa the beautiful *M. bullockoides* replaces its western and eastern ally. Our new *M. frenatus* appears to be nowhere commoner than in Sennar, where it is said to occur numerously on the banks of the White Nile and even more so on the Blue Nile (Vierthaler, Brehm). Rüppell first speaks of it as being of occasional occurrence in the north-eastern valleys of Abyssinia (the so-called Kulla). In Kordofan it must be a very rare species, as the industrious collector Petherick never met with it."

After a careful examination of all available specimens I cannot recognize Dr. Hartlaub's species as a valid one, and fully agree with Von Heuglin that the distinctions given are not constant, inasmuch as I possess in my own collection specimens from Senegal which agree closely with both so-called *M. frenatus* and *M. bullocki*. In one specimen from Senegal the frontal line and the blue margin to the black cheek-patch are as clearly defined as in the specimen from Sennar ; in the second these characters are but slightly defined, and in the third they are wanting. In both the specimens in the British Museum from Shonga there is no trace either of the frontal line or the blue margin to the black cheek-patch. It appears to me that these characters, on which Dr. Hartlaub proposed to separate the eastern from the western form, are signs of age, and that those examples which have the frontal line and the blue margin to the black eye-patch are fully

adult or very old birds. Referring to this question, Von Heuglin says (Orn. N.O.-Afr. i. p. 203) :—
" Central-Abyssinian birds are, as a rule, larger than East-African examples. Hartlaub and Finsch
separate the eastern *M. frenatus* from the western *M. bullockii*, as the latter never has the blue
edging to the black cheek-patches ; but this does not, in my opinion, constitute any specific
difference. According to Schlegel, only the very old males of *M. bullockii* have this blue border ;
and in any case there are specimens from East Africa in which it occurs only very slightly."

I find but little variation in the measurements of the various specimens I have examined, as
will be seen by the following table :—

	Culmen. in.	Wing. in.	Tail. in.	Tarsus. in.
Senegal (*Mus. H. E. Dresser*)	1·3	3·75	3·45	0·40
„ „	1·3	3·9	3·68	0·45
„ „	1·3	3·9	3·58	0·42
Gambia (*Mus. Shelley*)	...	3·9	3·70	0·40
„ „	1·4	3·9	3·60	0·42
Senegal (*Mus. Brit.*)	1·35	3·8	3·60	0·42
„ „	1·4	3·7	3·55	0·40
Shonga „	1·4	3·9	3·75	0·42
„ „	1·3	4·0	3·75	0·42
Senaar „	1·4	3·8	3·70	0·45

The specimen figured and described is from Senegal and is in my own collection.

In the preparation of the above article I have examined the following specimens :—

E Mus. H. E. Dresser.

a, b. Senegal, W. Africa (*Verreaux*).　*c.* Senegal (*R. B. Sharpe*).

E Mus. Brit.

a. River Gambia (*Whitely*).　*b, ♂.* Senegal (*Verreaux*).　*c.* Senaar (*Dr. Kotschy*).　*d, e.* Shonga (*Forbes*).

E Mus. Paris.

a. White Nile (*M. d'Arnaud*).　*b.* Senegal.

E Mus. G. E. Shelley.

a, b. Gambia (*Moloney*).

BOLESLAVSKIS HET. KATER
MELITTOPHAGUS BOLESLAVSKII

MELITTOPHAGUS BOLESLAVSKII.

BOLESLAVSKI'S BEE-EATER.

Merops bullocki, Vierth. Naumannia, 1852, p. 41, partim (nec Vieill.).

Merops boleslavskii, Von Pelz. Sitzungsb. k. Ak. Wiss. Wien, xxxi. p. 320 (1858) ; Hartl. J. f. O. 1861, p. 107 ; Heuglin, Orn. N.O.-Afr. i. p. 206 (1869).

Spheconax boleslavskii (Von Pelz.), Cab. Mus. Hein. ii. p. 133 (1859).

Merops frenatus, Heugl. J. f. O. 1864, p. 335, partim (nec Hartl.).

Merops boleslavskyi (Von Pelz.), Heuglin, J. f. O. 1864, p. 336.

Coccolaryna boleslavskii (Von Pelz.), G. R. Gray, Hand-l. of B. i. p. 101, no. 1227 (1869).

Figura nulla.

Hab. Northern Africa.

Ad. M. bullocki persimilis, sed gulâ flavâ nec rubrâ.

Adult (Casamance).—Closely resembles *Melittophagus bullocki*, except that the throat, instead of being red, is rich bright yellow. Total length 7·25 inches, culmen 1·5, wing 3·9, tail 4·05, tarsus 0·5.

It is with considerable hesitation that I include the present bird as a distinct species, and it is more from a lack of sufficient material to form a decided opinion on the subject, than from a conviction that it is clearly distinct, that I am compelled to do so. I have, indeed, only been able to examine a single specimen, now in the British Museum ; for the bird is so rare that I cannot find another in England. At first I thought it might prove to be an immature example of *Melittophagus bullocki* ; but the freshness of the plumage makes it improbable that this will be the case, and I also find that in the young of *Melittophagus gularis* the red throat-patch is assumed at once, and is not at first yellow, so that in all probability it will be the same as regards *Melittophagus bullocki*. It is, however, quite possible that the present bird may be a peculiar race of *M. bullocki* ; and I trust that collectors who read this article will use their utmost endeavours to obtain the necessary materials to fully elucidate this moot question.

The range of the present species appears to be about the same as that of *Melittophagus bullocki*, with which it has been found consorting. Von Heuglin states (*l. c.*) that he frequently obtained it with *Melittophagus bullocki* in Southern and Eastern Senaar, and looked on it as being the young of that species, as it agrees with it in size and (excepting in the colour of the throat) in plumage, and he also remarked that in various specimens the throat varied from brownish yellow and reddish brown to yellowish; but he adds that the fact of the freshness of its plumage is in favour of its validity as a distinct species, and he accordingly treats it as specifically separable from *Melittophagus bullocki*.

But little is on record respecting this bird. Dr. Vierthaler met with one in company with *Melittophagus bullocki* above Khartoum, and, as above stated, Von Heuglin records it from Southern and Eastern Sennaar.

In habits, and doubtless also in nidification, the present species does not differ from *Melittophagus bullocki*.

In the preparation of the above article I have only been able to examine a single specimen, the bird above described and figured, which is in the British-Museum collection.

REVOILS BEE-EATER

METELOPHAGUS REVOILI

MELITTOPHAGUS REVOILI.

REVOIL'S BEE-EATER.

Merops revoilii, Oustalet in Revoil's Faune et Flor. Çomalis, Ois. p. 5.
Melittophagus revoili (Onst.), Shelley, Ibis, 1885, p. 398.

Figura unica.
Oustalet in Revoil's Faune et Flor. Çomalis, Ois. pl. i.

HAB. Somali-land, Africa.

Ad. capite pallidè viridi pallidiore striato, plumis versus apicem cæruleo lavato : frontis lateribus et striâ superciliari cærulcis, striâ per oculum ductâ cum regione paroticâ nigris : nuchâ cum dorso supremo rufescenticervinis, plumis versus apicem viridi-cæruleo tinctis : dorso imo cum supra- et subcaudalibus cæruleis : alis viridibus, remigibus in pogonio interno et versus apicem fuscis : rectricibus viridibus vix cæruleo lavatis, omnibus (duabus centralibus exceptis) indistinctè rufescente terminatis : capitis lateribus, mento et gulâ albidis, gulâ imâ cum pectore rufescenti-cervinis vix viridi tinctis, subalaribus rufescenti-cervinis : rostro et pedibus nigris : iride fusco-rubrâ.

Adult (Somali-land).—Crown dull apple-green, most of the feathers tipped with pale greenish blue, giving the crown a striped appearance ; nape similarly coloured to the crown, but tinged with pale buff; upper back and upper surface of the wings dull parrot-green, most of the feathers slightly margined with pale blue ; all the quills except the elongated inner secondaries washed with blackish grey at the tip ; lower back light azure-blue, darkening to rich azure-blue on the upper tail-coverts ; tail dull green, the central rectrices darker and slightly washed with blue ; lores and a broad patch passing through and behind the eye black, margined above with azure-blue ; chin and throat white, becoming pale buff on the lower throat ; rest of the underparts and under wing-coverts brownish or golden buff ; under tail-coverts rich azure-blue ; bill and legs black ; iris brownish red. Total length about 10 inches, culmen 1·25, wing 3·05, tail 2·85, tarsus 0·42.

The second specimen from Somali-land is slightly richer in colour, and measures—culmen 1·2, wing 2·9, tail 2·8, tarsus 0·4.

WHEN I first read the description and saw the plate of the present species, I must confess that I was somewhat puzzled and in doubt as to whether it was not the immature bird of some already known species, and I deferred from time to time penning this article, trusting that I might have an opportunity of examining the type specimen; but continual calls on my time forbade my undertaking a trip to Paris in order to do so; and, besides, I was unable to send over my artist, Mr. Keulemans, in order that he might figure the type example for the present work, as his time was so fully occupied here.

However, I have now fortunately been able to examine two specimens obtained by Mr. E. Lort Phillips on the plateau of the Somali country above Berbera, which were most courteously placed at my disposal, one of which I have figured, and am glad to find that it is a most distinct and easily recognizable species, nearest allied perhaps to *Melittophagus bullocki*, but differing very appreciably and constantly from that Bee-eater. First obtained by M. Revoil in Somali-land, and described by M. Oustalet from the single example obtained by that gentleman, it has now been again obtained in the same country by Mr. Lort Phillips, and I believe that these three specimens are the only ones at present known to ornithologists.

A plate of this Bee-eater was published in M. Revoil's work (*l. c.*), and the following is the original description by M. Oustalet in the same work:—" Vertice, alis caudaque viridibus, gula et pectore albidis, abdomine fulvescente; superciliis, lumbis caudæque tectricibus superioribus et inferioribus cæruleis, regione interscapulari fulva, plaga postoculari, rostro pedibusque nigris."

According to Mr. Lort Phillips, this Bee-eater is common on the extensive plateau between Berbera and the Haynes River in Somali-land; and M. Revoil, who also met with it in the Somali country, states that it was observed there in considerable numbers, but he only brought back one specimen, the type of the species. Beyond these meagre details nothing whatever appears to be known respecting this rare Bee-eater, and the only specimens I have been able to examine are the following, one of which is the specimen I have figured and described, viz. :—

E Mus. E. Lort Phillips.

a, b. Somali-land (*E. L. Phillips*).

INDEX OF SPECIFIC NAMES

ADOPTED AND REFERRED TO.

DATES OF PUBLICATION.

Part I. Containing pages 1–40	. . . .	May 1884.
„ II.	„	„ 41–70	. . . .	July 1884.
„ III.	„	„ 71–106	. . . .	October 1884.
„ IV.	„	„ 107–132	. . . .	April 1885.
„ V.	„	„ 40 A–40 B, 133–144, Preface, Introduction, &c.	. . .	March 1886.

PRINTED BY TAYLOR AND FRANCIS, RED LION COURT, FLEET STREET.

www.ingramcontent.com/pod-product-compliance
Lightning Source LLC
Chambersburg PA
CBHW021530210326
41599CB00012B/1440